T0136863

ON RULES
REGARDING THE PRACTICAL PART
OF THE MEDICAL ART

◆

THE MEDICAL WORKS OF MOSES MAIMONIDES

Medical Aphorisms: Treatises 1–5

Medical Aphorisms: Treatises 6–9

Medical Aphorisms: Treatises 10–15

On Asthma

On Asthma, Volume 2

On Hemorrhoids

On Poisons and the Protection against Lethal Drugs

On Rules Regarding the Practical Part of the Medical Art

◆ ◆ ◆

Forthcoming Titles

Commentary on Hippocrates' Aphorisms

Medical Aphorisms: Treatises 16–21

Medical Aphorisms: Treatises 22–25

Medical Aphorisms: Hebrew Translations
by Nathan ha-Meʾati and Zeraḥyah Ḥen

Medical Aphorisms: Indexes and Glossaries

On Coitus

On the Elucidation of Some Symptoms and the
Response to Them (On the Causes of Symptoms)

The Regimen of Health

Maimonides

On Rules Regarding the Practical Part of the Medical Art

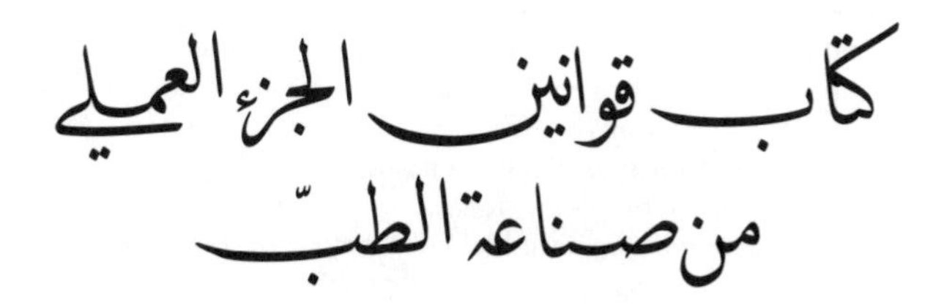

A parallel Arabic-English edition by

Gerrit Bos and Y. Tzvi Langermann

◆ ◆ ◆

PART OF THE MEDICAL WORKS
OF MOSES MAIMONIDES

Brigham Young University Press ◆ Provo, Utah ◆ 2014

Library of Congress Cataloging-in-Publication Data

Maimonides, Moses, 1135–1204, author.
On rules regarding the practical part of the medical art =
Kitab qawanin al-juz' al-amali min sina'a al-tibb : a parallel Arabic-English edition /
Moses Maimonides ; [edited by] Gerrit Bos and Y. Tzvi Langermann. — First edition.
p. ; cm. — (Medical works of Moses Maimonides)
Parallel title: Kitab qawanin al-juz' al-amali min sina'a al-tibb
Includes bibliographical references and index.
Summary: "A heretofore unrecognized Maimonidean medical text with parallel
Arabic-English translation covering a variety of topics including the suturing of
abdominal wounds"—Provided by publisher.
In Arabic and English.
ISBN 978-0-8425-2837-5 (hardback)
I. Bos, Gerrit, 1948– editor, translator. II. Langermann, Y. Tzvi., editor.
III. Maimonides, Moses, 1135–1204. Kitab qawanin al-juz' al-amali min sina'a
al-tibb. IV. Maimonides, Moses, 1135–1204. Kitab qawanin al-juz' al-amali min
sina'a al-tibb. English. V. Title. VI. Title: Kitab qawanin al-juz' al-amali min sina'a
al-tibb. VII. Series: Maimonides, Moses, 1135–1204. Medical works of Moses Maimonides.
DNLM: 1. Medicine, Arabic—history. 2. Aphorisms and Proverbs. WZ 290]

R733
610—dc23

2013022961

Printed in the United States of America.

First Edition

Contents

Sigla and Abbreviations

< >	supplied by editor, in Arabic text
[ ]	supplied by translator, in English text
[...]	illegible, in English text
(!)	corrupt reading
(?)	doubtful reading
add.	added in
om.	omitted in
inv.	inverted in
ditt.	dittography
emend. ed.	emended by the editors

Transliteration and Citation Style

Transliterations from Arabic and Hebrew follow the romanization tables established by the American Library Association and the Library of Congress (*ALA-LC Romanization Tables: Transliteration Schemes for Non-Roman Scripts*. Compiled and edited by Randall K. Barry. Washington, DC: Library of Congress, 1997; available online at www.loc.gov/catdir/cpso/roman.html).

Passages from *On Rules* are referenced by section number. Maimonides' introduction is designated as section 0.

Foreword to the Series

Brigham Young University and its Middle Eastern Texts Initiative are pleased to sponsor and publish the Medical Works of Moses Maimonides. The texts that appear in this series are among the cultural treasures of the world, representing as they do the medieval efflorescence of Arabic-Islamic civilization—a civilization in which works of impressive intellectual stature were composed not only by Muslims but also by Christians, Jews, and others in a quest for knowledge that transcended religious and ethnic boundaries. Together they not only preserved the best of Greek thought but enhanced it, added to it, and built upon it a corpus of scientific and philosophical understanding that is properly the inheritance of all the peoples of the world.

As an institution of The Church of Jesus Christ of Latter-day Saints, Brigham Young University is honored to collaborate with Gerrit Bos and other members of the academic community in bringing this series to fruition, making these texts available to many for the first time. In doing so, we at the Middle Eastern Texts Initiative hope to serve our fellow human beings of all creeds and cultures. We also follow the admonition of our own religious tradition, to "seek . . . out of the best books words of wisdom," believing, indeed, that "the glory of God is intelligence."

—Daniel C. Peterson
—D. Morgan Davis

Preface

We are very pleased to offer to the reader this edition and translation of Maimonides' treatise *On Rules Regarding the Practical Part of the Medical Art*. This treatise, extant as item 9 in a unique MS DCI of the Biblioteca Nacional de Madrid (formerly Escorial 888), fols. 109a–123a, is not mentioned in the recent bio-bibliographical literature. The great Jewish bibliographer Moritz Steinschneider identified the text in this manuscript as another copy of the treatise *On Asthma* and considered the title to be fictitious. However, from the introduction it is clear that it was written for the same patient for whom Maimonides wrote *On Asthma* and was meant to be part of that tract, although ill health prevented Maimonides from completing it at the time. He did eventually finish it, producing the text that we publish here. Maimonides' *Rules* is an independent, authentic work that fits the mold of his medical writings. It is written in the *fuṣūl* format—short, self-standing paragraphs that Maimonides wrote down from time to time, later to be organized into a series of monographs and one large book, his *Medical Aphorisms*. However, there is a basic difference between some *fuṣūl* in this work and those found in *Aphorisms*. Here, some represent a kind of itemized account, consisting of numbered lists of medical issues far more comprehensive than those in *Aphorisms*. Thus, one is left with the impression that they are the result of a life of learning and practice. Especially noteworthy is the unique *fuṣūl* dealing with surgery performed on serious abdominal wounds, as it seems to reflect Maimonides' experience with battlefield casualties.

This new edition is part of an ongoing project to critically edit Maimonides' medical works that have not been edited at all or have been edited in unreliable editions. This project started in 1995 at the University College London with the support of the Wellcome Trust, and now is proceeding at the Martin Buber Institute for Jewish Studies at the University of Cologne with the financial support of the Deutsche

Forschungsgemeinschaft. So far it has resulted in the publication of critical editions of Maimonides' *On Asthma* (2 vols.), *Medical Aphorisms* 1–15 (3 vols.), *On Poisons and the Protection against Lethal Drugs*, and *On Hemorrhoids*.

The series is published by the Middle Eastern Texts Initiative at Brigham Young University's Neal A. Maxwell Institute for Religious Scholarship. On this occasion we thank Professor Daniel C. Peterson, under whose direction this series has been prepared for publication, and his colleague, Dr. D. Morgan Davis, for their enthusiastic support of the project and dedication to it. Thanks are also due to Muhammad S. Eissa, Angela C. Barrionuevo, Andrew Heiss, Elizabeth Watkins, Felix Hedderich, Don Brugger, and David Calabro for their diligent editing, proofreading, and typesetting.

Introduction

Biography and Medical Works

Moses Maimonides, known under his Arab name, Abū ᶜImrān Mūsā ibn ᶜUbayd Allāh ibn Maymūn, and his Jewish name, Moshe ben Maimon, was not only one of the greatest Jewish philosophers and experts in Jewish law (halakah),[1] but an eminent physician as well. Born in Córdoba in 1138,[2] he was forced to leave his native city at age thirteen because of persecutions by the fanatical Muslim regime known as the Almohads and the policy of religious intolerance adopted by them.[3] After a sojourn of about twelve years in southern Spain, the family moved to Fez in the Maghreb. Some years later—probably around 1165—they moved again

1. For Maimonides' biographical data see *Encyclopaedia of Islam*, s.v. "Ibn Maymūn"; *Encyclopaedia Judaica*, s.v. "Maimonides, Moses"; Lewis, "Maimonides, Lionheart and Saladin"; Goitein, "Ḥayyē ha-Rambam"; Goitein, "Moses Maimonides, Man of Action"; Shailat, *Iggerot ha-Rambam*, 1:19–21; Cohen, "Maimonides' Egypt"; Ben-Sasson, "Maimonides in Egypt"; Levinger, "Was Maimonides 'Rais al-Yahud'?"; Davidson, "Maimonides' Putative Position"; Kraemer, "Life of Moses ben Maimon"; Kraemer, "Maimonides' Intellectual Milieu." For a fundamental discussion of all the available data concerning Maimonides' biography, see Davidson, *Moses Maimonides: The Man*, 3–74; and Kraemer, *Maimonides: One of Civilization's Greatest Minds*. For Maimonides' training and activity as a physician, see introduction to Maimonides, *On Asthma*, ed. Bos, xxv–xxx, and Bos, "Maimonides' Medical Works and Their Contributions."

2. While traditionally his date of birth is set at 1135, in the colophon to his *Commentary on the Mishnah*, Maimonides himself wrote in 1168 that he was then in Egypt and thirty years old. Goitein, "Moses Maimonides, Man of Action," 155, argues on the basis of this that the actual year of his birth should be put at 1138; see as well Leibowitz, "Maimonides: Der Mann und sein Werk," 75–76.

3. Following Graetz, *Geschichte der Juden*, 6:265, it is generally assumed that the family left Córdoba in the year 1148, when the city was conquered by the Almohads. Accordingly, Maimonides was ten.

because of the persecutions of the Jews in the Maghreb, this time going to Palestine. After staying some months there, the family moved on to Egypt and settled in Fusṭāṭ, the ancient part of Cairo.

It was in Cairo that Maimonides started to practice and teach medicine in addition to pursuing his commercial activities in the India trade.[4] He became physician to al-Qāḍī al-Fāḍil, Saladin's chief administrator, who brought many physicians to serve the sultan and his royal entourage.[5] Later, he became court physician to the sultan al-Malik al-Afḍal, after the latter's ascension to the throne in the winter of 1198–99. It is generally assumed that Maimonides died in 1204. The theory that for some years he served as *raʾīs*, or head, of the Jewish community is disputed. While Davidson argues against it,[6] Friedman argues in favor of it;[7] and according to Kraemer, Maimonides did serve as head of the Jews from September 1171 until 1173, in order to secure for the Jewish communities a favorable position with the Ayyubids, who had replaced the Fatimids as the ruling dynasty.[8] According to some sources, he served a second time in the 1190s, possibly between 1198 and 1199.[9]

Maimonides was a prolific author in the field of medicine, composing ten works considered authentic.[10] These works include the following major compositions: *Sharḥ fuṣūl Abuqrāṭ* (*Commentary on Hippocrates' Aphorisms*), *Kitāb al-sumūm wa al-taḥarruz min al-adwiya al-qattāla* (*On Poisons and the Protection against Lethal Drugs*), *Kitāb al-fuṣūl fī al-ṭibb*

4. Goitein ("Moses Maimonides, Man of Action," 163) has shown that Maimonides was already involved in this trade before his younger brother David perished in a shipwreck in 1169, and that he still had a hand in it in 1191, when he was practicing as a physician.

5. See Kraemer, *Maimonides: One of Civilization's Greatest Minds*, 215. See also *Encyclopaedia of Islam*, new ed., s.v. "al-Ḳāḍī-al-Fāḍil."

6. See Davidson, "Maimonides' Putative Position."

7. See Friedman, "Ha-Rambam 'Raʾis al-Yahud.'"

8. See Kraemer, *Maimonides: One of Civilization's Greatest Minds*, 191–92, 222–26.

9. Kraemer, *Maimonides: One of Civilization's Greatest Minds*, 227.

10. For his medical works see Meyerhof, "Medical Work," 265–99; Friedenwald, *Jews and Medicine*, 1:200–216; Baron, *Social and Religious History of the Jews*, 8:259–62; Ullmann, *Medizin im Islam*, 167–69; *Encyclopaedia Judaica*, s.v. "Maimonides, Moses"; Avishur, *Shivḥe ha-Rambam*, 33–36; Ackermann, "Ärztliche Tätigkeit"; Bos, ed., of Maimonides, *On Asthma*, xxxi–xxxii; and Langermann, "L'oeuvre médicale de Maïmonide." For a survey of editors and translators of Maimonides' medical works, see Dienstag, "Translators and Editors," 95–135; for Muntner's activity, see especially pp. 116–21.

(*Medical Aphorisms*), and *Mukhtaṣarāt li-kutub Jālīnūs* (*Compendia from the Works of Galen*).

The final six treatises are considered minor works: *Kitāb fī al-jimāᶜ* (*On Coitus*), probably written in 1190 or 1191 at the request of an anonymous high-ranking client; *Fī tadbīr al-ṣiḥḥa* (*On the Regimen of Health*), written at the request of al-Malik al-Afḍal; *Maqāla fī bayān al-aᶜrāḍ wa-al-jawāb ᶜanhā* (*On the Elucidation of Some Symptoms and the Response to Them*),[11] probably written after 1198 for the same al-Malik al-Afḍal when his condition did not improve; *Sharḥ asmāʾ al-ᶜuqqār* (*Commentary on the Names of Drugs*); *Risāla fī al-bawāsīr* (*On Hemorrhoids*); and *Maqāla fī al-rabw* (*On Asthma*).

Kitāb qawānīn al-juzʾ al-ᶜamalī min ṣināᶜa al-ṭibb (*Treatise on Rules Regarding the Practical Part of the Medical Art*)

Besides these ten works currently found in the bio-bibliographical literature, Maimonides is the author of the *Kitāb qawānīn al-juzʾ al-ᶜamalī min ṣināᶜa al-ṭibb* (*Treatise on Rules Regarding the Practical Part of the Medical Art*), which is extant in a unique manuscript, Madrid DCI (formerly Escorial 888), item 9, fols. 109a–123a. Steinschneider identified the text in this manuscript as another copy of the treatise *On Asthma* and considered the title to be fictitious.[12] Upon inspection, however, it is clear that *On Rules* is an independent, authentic work composed by Maimonides and that it fits the mold of Maimonides' medical writings. It is written in the *fuṣūl* (aphorisms) format: short, self-standing paragraphs that Maimonides wrote down from time to time, which he would later organize into a series of monographs and one large book, his *Medical Aphorisms.* As Maimonides tells us in the introduction to that book, this process was the custom of medical writers; indeed, the most detailed account of the *fuṣūl* as a literary genre is found in the introductory essay to Maimonides' *Medical Aphorisms.* In the course of years of studying and practicing medicine, Maimonides filled his notebooks with hundreds of *fuṣūl*, which formed the raw material for his medical writings. This method of composition allowed Maimonides, who was pressed for time and suffering from ill health, to avail himself of the same *faṣl* in more than one writing.[13] Some sixteen *fuṣūl* found in the present essay were included, with somewhat different wording, in other Maimonidean texts,

11. Traditionally titled *On the Causes of Symptoms.*

12. See Steinschneider, *Hebräischen Übersetzungen des Mittelalters*, 767.

13. Indeed, Maimonides is likely to have followed the same practice in the composition of his *Guide of the Perplexed*; see Langermann, "Fūṣūl Mūsā."

mainly his *Medical Aphorisms* (see the supplement at the back of the book). However, there is a basic difference between some *fuṣūl* in this work and those found in *Aphorisms*. Here these *fuṣūl* are a kind of itemized account, consisting of numbered lists of medical issues far more comprehensive than those in *Aphorisms*, leaving one the impression that they are the result of a life of learning and practice.

In a brief introductory paragraph, Maimonides reveals a few more details about our treatise. It was written for the same patient for whom Maimonides wrote his monograph *On Asthma* and was meant to be part of that tract. Ill health prevented Maimonides from completing it at the time; but he did eventually finish it, producing the text that we publish here. We know that it is a finished product because it closes with an admonition to pay close attention to the advice given in the treatise. It also gives some other interesting pieces of advice and a reference to the treatise *On Purgatives* found in book 13 of his *Medical Aphorisms*:[14]

> (86) One should set one's mind on all the things discussed in this treatise, compare them with one another, and then act accordingly. Beware of fear and of not being generous in [any matter, whether] small or large, and do not be excessive in something that the contemporary physicians consider to be needless. Do not treat evil diseases [so that] you will not be called a "physician of evil." Do not deviate from the course pursued by physicians [in general], and do not try a medicine that has not been tried before [and found to be] safe. I have mentioned all the precautions one should take in this regard in the treatise *On Purgatives*. This is what I intended [to write], praise be to God.

From a literary point of view, the passage just cited is clearly an ending. In fact, the thirteenth, and final, section of *On Asthma* contains some general rules for the preservation of health, over and beyond the specific advice that Maimonides offered to his asthmatic patient; this may be some vestige of Maimonides' original plan, which would have been that *On Asthma* should end with a comprehensive treatment of medical care. Unfortunately, we do not know the name of the person for whom *On Asthma* was written, although he was an important individual. We are also uninformed about the date of its composition. Hence we remain in the dark concerning the date and patron (client?) of *On Rules*. However,

14. Maimonides, *Medical Aphorisms*, chapter 13 (ed. and trans. Bos, 40–51).

it is very safe to say that *On Rules* was written very near the end of Maimonides' life. Maimonides did not begin to write on medicine until he was well advanced in years, and the completion of this monograph had to wait for some (unspecified) time after he finished *On Asthma.*

The Manuscript

The unique copy of our treatise is found in manuscript DCI of the Biblioteca Nacional de Madrid. The codex was dubbed "Medicina Castellana" by Guillén Robles, who gave a fairly detailed account of its contents.[15] One hundred and twenty years have passed since the publication of his catalogue and, as far as we know, no one has studied the codex since his time. The present study is limited to Maimonides' text. It is written in Arabic letters, in a Maghrebi hand, and was completed at the beginning of October 1424. The ink has faded; and the writing is difficult, in some places next to impossible, to decipher. The reader encounters not only many of the grammatical oversights common in texts of this sort, but also a few more serious textual problems. We have been as judicious as possible in suggesting emendations to the text, but in some cases it was necessary to emend in order to make the text coherent; and in a very few instances, we have no satisfactory solution to offer.

What Does Maimonides Mean by "Practical Medicine"?

Medicine had traditionally been divided into two distinct subdisciplines, the theoretical and the practical. Maimonides' *On Rules* clearly belongs to the latter. However, the precise definition of this category was a matter of some discussion and controversy. Ibn Sīnā insisted that the practical part of medicine was no less a science than the theoretical part; it too deserved to be called *ᶜilm*. According to him, the practical part develops the general, scientific rules that guide practice; it is not simply a list of practical instructions. He knows quite well that in defining the practical side in this manner, he is breaking ranks with many, perhaps most, of the physicians of his day. In particular, it seems that Ibn Sīnā is targeting Ḥunayn ibn Isḥāq, whose very influential *Mudkhal* (*Introduction*, what the Greeks would have called *Eisagoge*) opens with the proclamation that there are two parts to medicine: *naẓar*, or "science"; and *ᶜamal*, or "practice." A simplistic understanding of this classification had taken

15. Guillén Robles, *Catálogo de los manuscritos árabes*, 246–48.

hold, which understanding Ibn Sīnā rejects outright.[16] He clarifies his own position in the very first chapter of his monumental *Qānūn*. His remarks will be very useful for us in assessing Maimonides' treatise, so we will cite them at length:

> You ought not to hold the opinion that their intention [in dividing medicine into these two divisions] was that one of the divisions of medicine studies science, while the other division comprises directions for practice [*al-mubāshara li-l-ᶜamal*], which is where the fantasy of many investigators leads them concerning this topic. Instead, it is your duty to know that the intention here is something else, namely, that each of the two divisions is nothing other than a science [*ᶜilm*]. However, one of them concerns the science of the principles of medicine, and the other, the science of how it is to be applied [*kayfiyyat mubāsharatihi*]. Further, the first section is characterized in particular by the name "science" or by the name "investigation" [*naẓar*; may also be translated "speculation"], while the other is characterized in particular by the name "practice" [*ᶜamal*]. "Investigation" refers to that whose study yields [correct] belief alone, without concern to clarify how this may be applied. Examples of this are the statement, made in medicine, that there are three types of fever and nine temperaments. "Practice" does not refer to actual practice, nor to the application of bodily movements. Instead, it is the part of medical science whose study yields an informed opinion [*rayʾ*] that links to the clarification of how to practice. For example, it is stated in medicine with regard to hot tumors [*awrām ḥāra*], that, at their inception, the treatment ought to impede them, cool them, and bring them into the open. After that, the [impeding drugs][17] should be mixed with softeners. Then, after the limit has been reached, until they begin to shrink, the solvent softeners should be reduced; but this [course of treatment] should be given in the case of tumors that are caused by materials that have been expelled by primary organs. Studying this yields an informed opinion [suggestion], which clarifies how the practice should be executed.[18]

16. It would be interesting to see what, if anything, Ibn Sīnā has to say in his gloss on this passage. Unfortunately, his commentary to the *Mudkhal* has never been studied, though many copies exist; see Anawati, *Essai de bibliographie avicenienne*, no. 144. In fact, the *Mudkhal* itself has been sorely neglected. The copy consulted here is MS Vatican 348, one of several copies of the Arabic in Hebrew letters.

17. Lit., impediment.

18. Ibn Sīnā, *Kitāb al-qānūn fī al-ṭibb*, 1:3.

Maimonides identified himself strongly with the western (Maghrebi/Andalusian) tradition in medicine, and, as such, he would not have been beholden to Ibn Sīnā.[19] He does not directly address the definition of the practical part of medicine in the treatise that we publish here. However, inspection of the *fuṣūl* that make up the treatise shows that, by and large, he agrees with Ibn Sīnā's approach. Some of the *fuṣūl*, especially at the beginning of this collection, give precisely the type of general rules that Ibn Sīnā spoke of. A good example is *faṣl* 18:

> (18) One should evaluate the cause [of the ailment], the ailment [itself], and its symptoms, as each one stands in relation to the others; then one should pay attention to the [thing] that causes the strongest disturbance [in the body] and weakens its strength most of all. If you come across something that counteracts all these things together or that counteracts that which is most severe but is beneficial for the other things, you should rely on it. But if you do not come across something like that, you [should turn to treating] that which is most important without neglecting the other [things]. If the cause of the fever is alarming [very dangerous], one should take care to root it out, even if this [action] increases the heat of the fever. And sometimes the fever itself is so high and severe that one should hasten to cool it and extinguish it, even if this [treatment] worsens the cause. And in some cases the fever is accompanied by severe symptoms—such as a collapse of strength, excessive diarrhea, and fainting—and one should quickly pay attention to the symptom [first], so that it disappears, and then start treating the fever.

Maimonides here advises the physician to assess the relative urgency of three factors: the cause of the disease, which, in the case of fevers, was taken to be superfluous material that the body had not expelled; the ailment (the fever, or just how hot the body is); and the symptoms. Ideally, the physician would employ a remedy or treatment that addresses all three. However, in general this will not be the case; moreover, for a given patient, one of the three factors may pose an immediate threat and demand a solution, even if the solution exacerbates one of the other two. This would imply, for example, that if the patient is extremely weak, he should take some food. This contradicts the opinion, generally held at

19. See Langermann, "*L'oeuvre médicale de Maïmonide*," 281–82. Concerning the disdain in which Ibn Sīnā was held by Maghrebian physicians, see *idem*, "Another Andalusian Revolt?," 366.

the time, and in fact endorsed by Maimonides in his own *Aphorisms* (10.68), that feverish individuals should not eat under any circumstances, since food is the main source of material superfluities that fuel the fever. The implication of *faṣl* 18 is in fact stated clearly at the end of *faṣl* 20, another example of practical medicine that matches the definition of Ibn Sīnā:

> (20) The issue of food [to be given to someone ill] has six aspects:
>
> i. One of these is the strength [of the patient]: If it is sound, then he can tolerate waiting and a lightening [of his diet]. The opposite [also holds true—if his strength is weak—then he must be given food right away, and his diet should not be lightened].
>
> ii. The length of the disease: If the disease lasts for a long time, the [patient's] strength should be maintained by means of food.
>
> iii. The disease [itself]: [The quality of the food given] should be opposite to [that of] the disease: For instance, in the case of fever caused by the putrefaction of thick, viscous humor, the food should thin out the thickness of the humor.
>
> iv. The time: [The food] should be administered at the usual time, both during [healthy periods] and when the attack abates.
>
> v. The digestive organs: If the stomach or the liver is affected by a tumor and food is administered before a [fever] attack, [the nourishment] is detrimental for the patient, especially when the body is overfilled. When one of [these organs] is weak because of a bad temperament or because of the influx of humor but it is not affected by a tumor, food is appropriate, even during a [fever] attack, especially when the body is not congested.
>
> vi. The magnitude of overfilling: When the body is overfilled, one should take less food; but when it is deficient, one should take more food, even during a fever attack.

We are left to wonder what this tells us about the development of Maimonides' thinking on the subject. Which text represents his latest thinking on the topic? The issue is not raised in the last of the twenty-five sections of Maimonides' *Aphorisms*, which contains his critique of Galen; but that last section was not edited (from Maimonides' notebooks, where these *fuṣūl* were recorded over the years) by Maimonides himself, but rather done posthumously by someone else.[20] What would Maimonides have done in his medical practice? Again, we do not know. However, another chapter from our text does preserve a very interesting and hitherto unknown remark by Maimonides about his own practice.

In section 33, Maimonides introduces his own observation, which essentially nullifies the rule given in textbooks. Galen had advised feverish patients to visit the bathhouse, and his opinion was repeated by "all physicians." Maimonides questions this practice—not directly contradicting Galen, but rather claiming that in his day, no one knows any longer about the effect of the bath upon fevers. In effect, Maimonides has thrown out Galen's rule, and instead advises treating fevers by diet or bleeding. This same chapter gives as well some advice concerning specific food substances. As such, it may possibly not be wholly in line with Ibn Sīnā's guidelines for practical medicine—guidelines that Maimonides never explicitly accepted, as we recall. Here is the passage:

> (33) When a fever occurs, one should examine it. If it is an ephemeral fever, the patient should enter the bathhouse when the fever abates, [according to] the opinion of all physicians. But I caution them against taking him to the bathhouse because we know little about this nowadays. And although there may be someone who knows about the nature of fevers and about the effect of [going to the bathhouse], we do not know about it [even] if we look into it, let alone knowing the effect. If [the patient suffers from] blood [fever (synochous fever)], one should hasten to bleed him. If [the fever originates from] yellow bile, the humor from which it originates should be evacuated using ingredients that have the property to cleanse and evacuate without heating, such as tamarind [*Tamarindus indica*], pears, pomegranate juice, oxymel, barley groats, and spinach. If [the fever] is chronic, one should either, in the beginning, give the patient ingredients that refine the coarseness of the humor [that causes that fever], thin its viscosity, and help its concoction; or, in the end, when the humor is concocted, give him ingredients that evacuate it. If [the patient suffers from] hectic fever, one should take care to cool and moisten [his body] and to revive his vigor.

Most of the advice given in this treatise is found in one form or another in the medical literature of the day, if not in Maimonides' own

---

20. See translator's introduction to Maimonides, *Medical Aphorisms* (ed. and trans. Bos), 1:xx–xxix.

writings. However, his procedure for treating serious abdominal wounds in which the intestines have become dislodged has not, to the best of our knowledge, been described in published studies of medieval texts. Different instructions for procedure, known as gastrorrhaphy (literally "suturing the abdomen") are given by Celsus and Galen, among others. Maimonides describes it in this way:

> (80) If the omentum or intestines protrude, raise the patient by his hands and feet, in hot air and in a manner such that the abdomen is drawn upward and becomes clearly visible, while the organ does not become cold. Then the patient should be softly shaken and gently put to sleep in this position on a flat bed raised at its extremities. Once the patient has been put to sleep and tied up in the middle [?], one should make efforts to return [the omentum or intestine] inside [the body]. Then one should draw the edges of the wound together and cut it carefully. If the patient needs to defecate, give him a clyster with [ingredients] that expel the feces, and alleviate the pain with astringent black wine or the like.
>
> (81) If one needs [to] suture [the wound], one should bring both edges of the wound together and stitch them, then fasten the thread with a double knot and cut it off. Then skip over a small [part?] of the wound, join the two edges, and stitch them in the same way. Continue to operate in this manner until you reach the end of the wound. The threads should consist of [material] that does not decay quickly.

These paragraphs raise many interesting issues that we can touch upon only very briefly. First, with regard to the prescribed treatment: Raising the hips, shaking the patient gently so that the intestinal coils return to their original position, and using sedatives or anesthesia are recorded by Celsus and others. The ancient authorities disagree, however, on the method for suturing, and here Maimonides' advice differs from them all.[21] Celsus recommends two close rows of stitches; Galen advises suturing the peritoneum to the abdominal wall, or—this is the preferred method—stitching the peritoneum to the peritoneum and the abdominal wall to the abdominal wall. Maimonides does not specify

---

21. See Papavramidou and Christopoulou-Aletra, "Ancient Technique of 'Gastrorrhaphy.'" Latin physicians of the medieval period differed as to whether wounds in which the intestines protrude are necessarily fatal; some, however, report in detail cases that were treated successfully; see McVaugh, *Rational Surgery*, 114–77.

which structures are to be sutured, but (if we understand him properly), he advises a series of single stitches, each secured with a double knot, after which the thread is cut; the ancient authorities all describe continuous stitching. In fact, we find a set of instructions in Maimonides' epitome of Galen's *Art of Cure*, which we cite here in the translation of Uriel Barzel:

> If the intestine remains inflated and projecting, we must cut the inner skin (peritoneum) to the extent necessitated by the intestine. You must take care that the intestines will not rest on top of that part of the intestine that was projecting and will not press on it. If you concentrate on that you will know that if the wound is on the right side you should order the patient to turn on his left, and if he is on the left side you should order him to turn on his right. Thus you must always aim to have the wounded side above the other one. You must also cover the entire area of the wound with a "barrier" from the outside, to join its parts, and then to expose it gradually upon sewing it until the entire wound is sewed well, in the same manner as it is done by some doctors who bring together each part to its naturally corresponding part, joining the edge of the inner skin to the other one, and the edge of the soft part to its other edge. This is a better method than the common stitch in which all four edges, those of the soft part and those of the inner skin, are joined by one stitch.[22]

We must recall that the epitomes exhibit the exact words of Galen; our treatise, by contrast, records Maimonides' own views. Of course, Maimonides cites the exact words of Galen in their Arabic versions, which, in some cases, differ significantly from the Greek. It would be interesting, perhaps as part of a study on Maimonides' views on surgery, to compare the Greek with the Arabic translations of the relevant passages and to bring into the discussion passages from other medieval physicians; such a task lies beyond the purview of the present undertaking.

Finally, just why did Maimonides choose to include this sort of procedure in this treatise? As it seems to us, abdominal wounds such as those described in the paragraphs in question are most likely to be incurred in battle. Maimonides certainly saw his share of bloodshed, beginning with the rampages of the Almohads that forced him, then a mere teenager, to flee from Andalusia. Nonetheless, our treatise was written towards the

---

22. Barzel, *Art of Cure*, 76–77. Maimonides' *Art of Cure* belongs to the series listed and described above, section 2, *Mukhtaṣarāt*.

end of his life, for a patron who was almost certainly connected to the Ayyubids, if not a member of that ruling dynasty. Legend has Maimonides putting his medical knowledge to use in the wars against the Crusaders,[23] and indeed he praises al-Qāḍī al-Fāḍil, the power behind Saladin, for scourging the infidels.[24] Perhaps, then, the particular rules for practical medicine found in the paragraphs cited above were written for those who treated battlefield casualties.

At the end of our discussion of the main themes of *On Rules*, we would like to conclude by expressing our gratitude in discovering a new treatise by Maimonides. It is with great pleasure that we present here a welcome addition to his medical opera, and one that informs us about some new dimensions to his medical expertise.

23. See Kraemer, *Maimonides: One of Civilization's Greatest Minds*, 160–61.

24. Indeed, Maimonides' relationship with al-Qāḍī al-Fāḍil is a very interesting topic that deserves a study of its own: it should prove valuable not just for the biographies of these two important individuals, but also for the relationships between Jews and Muslims in general, the internal governance of the two communities, and issues of public health. For Maimonides' appraisal of the accomplishments of al-Qāḍī al-Fāḍil, see the dedicatory introduction in *On Poisons* 1–7 (ed. Bos, 1–8). Tzvi Langermann lectured on the relationship of Maimonides and al-Qāḍī at a symposium in honor of Professor Joel L. Kraemer held several years ago at the University of Chicago, and he hopes to publish his findings soon.

ON RULES REGARDING THE PRACTICAL PART OF THE MEDICAL ART

◆

The Treatise on Rules Regarding the Practical Part of the Medical Art

[composed] by the Head Abū ᶜImrān Mūsā ibn ᶜUbayd Allāh, the Israelite from Córdoba, may God have mercy on him

(0) He said: My honorable master, the Pillar of Faith (may God protect him), enjoined me to compose a treatise on rules regarding the practical part of medicine using concise aphorisms. I carried out his command (may God grant him lasting happiness), and I followed the path whose goal came [to fruition] in writing it. I had intended, at the time that I wrote the treatise on the [illness] my master was suffering from (asthma), to put these aphorisms together with what was [in the treatise *On Asthma*], but I was prevented from doing so by illness and [thus] did not carry out [my intention] at that time. Now, however, I will begin [to do so], God willing.

(1) The first thing to consider is the improvement of the air, then the improvement of the water, and [then the improvement of] foods. Know that one's body is composed of matters that are in a constant [state] of dissolution and fluidity. For this reason [the body] does not remain in exactly the same condition, but in conditions that resemble each other, and this is called health. [When the body] deviates from the correctly

(fol. 109a)

كتاب قوانين الجزء العملي من صناعة الطبّ

للرئيس أبو عمران موسى بن عبيد الله القرطبي الاسرائيلي رحمة الله عليه

قال: كانت حضرة سيدي وعماد ‹دي›ن حرسها الله قد أوجبتني أن أؤلّف

٥ لها كتابا في قوانين الجزء العملي من أجزاء الطبّ على طريق الفصول والاختصار فأمثلت أمرها أدام الله سعدها وسلكت الطريق الذي قصده جاء في كتبه وقد كنت عزمت. حين آلّفت المقالة في الشكاية التي كانت بمولاي وهو الربو أن أؤلّف هذا الفصول مع التي هناك ولاكن عاقني مرض و لم أكملها هناك فهنا مبتدئي بها إن شاء الله.

(١) فصل: أوّل ما ينبغي أن يعتنى بإصلاح الهواء وبعد ذلك بإصلاح الماء

١٠ والأغذية. واعلم أنّ أبداننا مركّبة من جواهر؟ دائمة التحلّل وسيلان وهي لذلك لا تستقرّ على حال واحدة بعينها ولاكن قد تبقى على أحوال متشابهة ويسمّى ذلك صحّة وقد

٣ الاسرائيلي] .emend. eds الاسرالي MS || ٤ قد أوجبتني] .emend. eds فأوجبتني MS

proportioned causes,[1] this is called disease. For this reason health should be preserved by means of those things that maintain the [correct] mutual proportion, while illness should be treated by things that oppose [the illness] until [the body] returns to a harmonious [condition].

(2) There are three things by which health is maintained: (i) replacing the innate heat that has dissolved from the body and its organs; (ii) evacuating the residues that have accumulated in the body; and (iii) taking care that the body is not quickly affected by old age. These three objectives can be realized if one puts before one's eyes the assessment of the quantities and qualities of the established factors [of health] to which the physicians pay attention.

(3) Pay attention every day to improving the air that reaches the body through inhalation so that it will be perfectly balanced and free from all that might pollute it. The finer the pneuma is, the more sensitive it is to alterations in the air.[2] The natural pneuma is coarser than the vital pneuma, while the vital [pneuma] is coarser than the psychical [pneuma].[3] Likewise, pay attention to the concoction of the bodily humors in the stomach, vessels, liver, and [other] organs. [Take care] to decrease [their quantity] when it is too large and to increase [their quantity] when it is too small, to refine [the humors] that are coarse, and to dilute [the humors] that are viscous, and to balance [these humors]. [Take care] to exchange these [unhealthy] materials with opposite ones, so that every single one becomes again as it should be.

تتغيّر عن الأسباب المتناسبة ويسمّى ذلك مرضا. ومن أجل ذلك يجب ضرورة أن يكون حفظ الصحّة بالأشياء التي تحفظ التناسب وأن يكون علاج المرض بالأشياء التي تضادّه إلى أن يعود للتشابه.

(٢) فصل: الأشياء التي تتقوّم بها الصحّة ثلاثة أحدها إخلاف بدل ما يتحلّل ٥ من البدن من الحرارة الغاريزية والأعضاء والثاني تنقية الفضول المجتمعة في البدن والثالث العناية بأن لا يسرع الهرم إلى البدن وإمكان وجود هذا الأغراض الثلثة يكون إذا جعل تقدير كميات وكيفيات الأسباب المسنّة التي ينظر فيها الأطبّاء نصب العين.

(٣) فصل: اعن في كلّ يوم بإصلاح الهواء الذي يرد (fol. 109b) البدن ١٠ بالتنفّس حتّى يكون في غاية الاعتدال والنقاء من كل شيء يدنّسه وكلّما كان الروح ألطف كان تغيّرها لتغيّر الهواء أكثر فالروح الطبيعي أغلظ من الحيواني والحيواني أغلظ من النفساني. وكذلك اعتنى بإنضاج أخلاط البدن في المعدة والعروق والكبد والأعضاء وبنقص ما كان منها زائدا والزيادة في ما كان منها ناقص وتلطيف ما كان غليظا وبقطع ما كان منها لزجا وتعديلها وكذلك بإبدال في أضداد هذا ١٥ الأشياء حتّى يردّ كلّ واحد منها على ما ينبغي.

٣ للتشابه] .emend. eds للتشبّه MS || ٦ وجود] .emend. eds اوجود MS || ٩ اعن ... النفساني] قال جالينوس واعن (واعني) بأمرجوهر الهواء الذي يرد البدن بالتنفّس حتّى يكون في غاية الاعتدال والنقاء من كلّ شيء يدنّسه. قال المؤلّف وكلّ ما كانت الروح ألطف كان تغيّرها لتغيّر الهواء أكثر فالروح الطبيعي أغلظ من الروح الحيواني والحيواني أغلظ من النفساني MS | *Fī tadbīr al-ṣiḥḥa* 4.1; *Maqāla fī al-rabwi* 13.2–3 || ١٣ وبنقص] .emend. eds وبنفط MS | ناقص] .emend. eds نافط MS

(4) Pay attention to the preservation of health through exercise, massage, bathing, and other means that strengthen the organs, expel their residues, and preserve their temperament. [This holds good] especially for the major organs and for those organs that carry out an activity in the body that is of major importance, such as the heart, liver, stomach, and brain. These kinds of organs need more attention than the other ones.

(5) There are two kinds of exercise: The first is [the kind] that is both exercise and art [work], such as agriculture and every art in which the body is put into action. The other is mere exercise, such as running, wrestling, and playing with the [small] ball.[4]

(6) One should exercise only after the food in the body has been digested and its residues expelled. Digestion in the stomach and intestines is indicated by the evacuation of excrements in a natural way, and in the liver and [its?] vessels, by the evacuation of the urine in a proper way.[5]

(7) Massage, or rubbing, should be [applied] judiciously; [in the beginning it should be done] little by little and softly, then increased slowly until one reaches the right quantity.[6] Every [massage] should be done with the hands moving in different directions so that not one spot remains without rubbing or massage. The right quantity also entails heating the body and concocting [the humors] in the organs. Massage should be applied either before exercise (and then it should be soft) or afterward (and then it should be hard). If one rubs the body in order to moisten it, one should do so after bathing in sweet, lukewarm water.

(8) Those whose health is deficient should exercise, bathe, and receive massage and rubbing in air contrary to their bodily condition.

(٤) فصل: استعن في حفظ الصحّة بالرياضة والدلك والاستحمام وسائر ما يقوّي الأعضاء ويخرج عنها فضولها ويحفظ عنها أمزجتها ولا سيّما ما كان منها رئيسيا وما كان فعله في البدن فعلا شريفا كالقلب والكبد والمعدة والدماغ فإنّ أمثال هذا الأعضاء ينبغي أن تصرف العناية إليها أكثر من غيرها.

٥ (٥) فصل: الرياضة صنفان: أحدهما رياضة وصناعة معا كالفلاحة وكلّ صناعة ينبعث فيها البدن والآخر رياضة فقط كالإحضار والصراع واللعب بالكرة.

(٦) فصل: إنّما ينبغي أن يستعمل الرياضة بعد انهضام الغذاء في البدن وخروج فضوله ويستدلّ على ذلك إمّا في الهضم الكائن في المعدة والأمعاء بأن يخرج البراز ١٠ على المجاري الطبيعية وإمّا في الكبد والعروق بأن يخرج البول على ما ينبغي.

(٧) فصل: الدلك والمرخ ينبغي أن يكون أثيرا وهما قليلا قليلا برفق ثمّ يزاد فيها قليلا قليلا إلى أن يبلغ المقدار القصد وينبغي أن يكون كلّ واحد منهما بأيدي تمرّ في جهات مختلفة لئلا (fol. 110a) يبقى موضع ‹لم› يلحقه المرخ والدلك والقصد المقدار فيها هو أيضا إسخان البدن وإنضاج ما في الأعضاء وينبغي أن يكون الدلك إمّا قبل ١٥ الرياضة فليّن وإمّا بعدها فصلب. وإذا كان الغرض في المرخ ترطيب البدن فينبغي أن يكون بعد الاستحمام بالماء العذب الفاتر.

(٨) فصل: الهواء الذي ينبغي أن يكون فيه الرياضة والدلك والاستحمام والتمرّخ أما في الأبدان التي صحّتها نقيصة فهو ضدّه للحال التي هي عليها.

---

١ استعن] .emend. eds استعين MS || ٩ إمّا] .emend. eds ما MS || ١٤ إسخان] .emend. eds أسخن MS

(9) Exercise is not safe for [those] bodily parts [that suffer from] a discharge of [superfluous] matters into them. On the contrary, one should let them rest. Those parts that are opposite to the [affected] parts—[namely, those parts] away from which the [superfluous] matters are diverted—should also rest.

(10) Children need that which strengthens their bodily parts and their functions and which prepares [these parts] for the function they are intended for when they reach puberty. [They also need] that which preserves the innate heat in them and expels only the residues. Young people [need] that which preserves the powers of their bodily parts and which moistens according to the surplus of dryness in them. They also [need] that which expels[7] the vaporous and other residues from them. The elderly and old people need that which heats and moistens commensurate to their tendency toward cold and dryness. They also [need] that which strengthens the faculties of their organs and which expels[8] the residues. One ought to do likewise for the seasons of the year, the other temperaments, foods, and so forth, so as to provide for every [case] that which is fitting for it.

(11) If a body is on the verge of putrefaction, it can endure hunger and thirst more easily than other bodies. One should protect one's body against putrefaction using ingredients that expel the bile and that have a cooling and drying effect. One should also be wary of things that cause putrefaction, such as extreme heat and indigestion. A bilious body cannot endure hunger, thirst, or excessive exercise. Such a body should be protected from becoming ill through moderate exercise and massage; bathing in sweet, lukewarm waters; and foods that produce pleasant vapors, cool, and moisten without viscosity. [Such foods are] barley broth, well-prepared bread,[9] chickens that lay eggs, the testicles of roosters,[10] the yolk from chicken eggs, soft-boiled eggs,[11] rockfish,[12] and every [fish] that does not have bones. In general, one should beware of everything that is sharp,[13] that obstructs the smoky vapors [from leaving the body],[14] and that closes the pores. One should also avoid

(٩) فصل: الأعضاء التي تنصبّ إليها الموادّ الرياضة لها غير مأمونة بل ينبغي أن تراح وتراح الأعضاء التي تقابلها على المحاذاة بتميّل الموادّ عنها.

(١٠) فصل: الصبيان يحتاجون إلى ما يقوّي الأعضاء والأفعال ويجعلها مستعدّة نحو الصناعة التي يقصد بهم إليها عند البلوغ وإلى ما يحفظ عليهم الحرارة
٥ الغريزية وينفض الفضول فقط والشباب إلى ما يحفظ قوى الأعضاء وإلى ما يرطّب بقدر اليبس الزائد فيهم وإلى ما ينفض الفضول الدخانية وغيرها منهم والكهول والمشائخ يحتاجون إلى ما يسخّن ويرطّب بقدر الميل فيهم إلى البرد واليبس وإلى ما يزيد في قوى الأعضاء وإلى ما ينفض الفضول. وكذلك ينبغي أن تفعل في فصول السنة وسائر المزاجات والأغذية وغيرها حتّى تجعل كلّ واحد منها موافقة.

١٠ (١١) فصل: البدن المستعدّ للعفن أصبر على الجوع والعطش من سائر الأبدان وينبغي أن تحفظ الأبدان من العفن بالأشياء التي تنفض المرّة وبالتي تبرّد وتجفّف وبالتحرّز من أسباب العفن كالحارّ المفرط والتخم. والبدن المراري ليس له صبر على الجوع ولا على العطش ولا على الرياضة المفرطة وينبغي أن (fol. 110b) تحفظ هذا الأبدان من ال‹و›قوع في الأمراض بالرياضة والدلك المعتدلين والاستحمام بالمياه
١٥ العذبة الفاطرة والأغذية التي تولّد البخارات العذبة وتبرّد وترطّب من غير غلظ كحسو الشعير والخبز المحكم الصنعة والفراريج البائضة وخصى الديوك وصفرة بيض الدجاج والنيمرشت والسمك الرضراضي وكلّ ما لا شوك فيه. ويحرز بالجملة كلّ حرّيف وكلّما

١٤ المعتدلين] .emend. eds المعتديلين MS || ١٥ وتبرّد] .emend. eds وتغليظ MS وتبريد MS[1] || ١٦ الصنعة] .emend. eds الصناعة MS | والفراريج البائضة] .emend. eds والفراريج الباض MS فولد الفراريج الباض MS[1] || ١٧ ويحرز] .emend. eds ويحزر MS

exertion,[15] anger, sleeplessness,[16] bathing in astringent waters, and the like. One should similarly take care of the other [types of] bodies.

(12) A body that is extremely fat should be made more lean by eating less food; [by eating] salty things; [by drinking] thinning beverages; [by taking] refining drugs, especially diuretics; by promoting perspiration through exercise; by bathing on an empty stomach; by sleeping before taking a meal; and by bathing in salty waters. A body that is extremely lean should be made more fat by eating abundant food; by taking sweet and fat things and thickening beverages; by bathing in sweet waters with a full stomach; and by sleeping after taking a meal. If one intends to increase the size of a fleshy part, one may do according to the story that is told—namely, that a coppersmith irritated part of his body by beating it lightly with branches and then rubbing pitch on it, and that he used to do so once every three or four days, and that, in a short time, he greatly increased the size of that part.[17]

(13) Obese bodies that have a hot and moist temperament and a hot and sweet vapor originating in them can endure the postponement of [the consumption of] food, because of the small quantity of smoky vapor that they contain. But lean bodies that have a hot and dry temperament and a sharp vapor originating in them cannot endure such a postponement. Bodies most unable to endure fasting are those that are cold and dry, because a dry body is emaciated by abstention from food.

(14) Health should be preserved and increased, while illness should be decreased and abolished. Health is preserved through the soundness of the organs, the pneuma,[18] and the innate heat.[19] This [can be achieved] through suitable nutrition, a quiet life, the generation of good blood and only few superfluities, and [by] improving breathing through [the inhalation of healthy] airs and winds.[20]

يحقن البخار الدخاني ويسدّ المسامّ ويحرز التعب والغضب والسهر والاستحمام بالمياه القابضة وما أشبه لذلك وعلى هذا السبيل ينبغي لك أن تربّي سائر الأبدان.

(١٢) فصل: البدن المفرط السمن ينبغي أن يهزل بتقليل الغذاء وأخذ المالح والشراب الرقيق والأدوية الملطّفة وخاصّة المدرّة للبول وبالتعريق بالرياضة والحمّام
٥ على الجوع والنوم قبل أخذ الطعام والاستحمام بالمياه المالحة. والبدن المفرط الهزال ينبغي أن يسمن بالتوسّع في الغذاء وأخذ الحلو والدسم والشراب الغليظ والاستحمام على الشبع بالمياه العذبة والنوم بعد أخذ الطعام. وإذا قصدت إلى زيادة مقدار أحد الأعضاء اللحمية فعلى هذا الذي ذكره جاء: كان بعض النخّاسين يلذع العضو بضرب خفيف بقضبان ثمّ يطليه بزفت وكان يفعل ذلك كلّ ثلاثة أو أربعة أيّام مرّة
١٠ وكان يعظم العضو في زمان يسير عظما كثيرا.

(١٣) فصل: الأبدان الخصيبة التي مزاجها حارّ رطب والبخار الذي متولّد فيها حارّ عذب تحتمل الصبر على الطعام لقلّة البخار الدخاني فيها والأبدان النحيفة التي مزاجها حارّ يابس والبخار المتولّد فيها حرّيف لا تحتمل الصبر على ذلك وأقلّ الأبدان احتمالا للصوم الباردة (fol. 111a) اليابسة لأنّ البدن اليابس ينهكه الإمساك
١٥ عن الغذاء.

(١٤) فصل: الصحّة ينبغي أن تحفظ وتزاد فيها والمرض ينبغي أن ينقص منه وأن يبطل. وحفظ الصحّة يكون بسلامة الأعضاء والروح والحرارة الغريزية وذلك بالتغذية الموافقة وقرار العيش وتولّد الدم الجيّد وبعض الفضول وإصلاح التنفّس بالأهوية والأرايح.

---

١ يحقن] emend. eds. يخضن MS | ويحرز] emend. eds. ويحزر MS || ٣ السمن] ما add. MS

(15) To apply oneself in acquiring knowledge of every single part of the body and its peculiarities is doubtlessly useful for therapy.[21] This is so because if, [for instance], the liver suffers from a hot tumor, it needs [a remedy] that opens [that tumor] and cleanses it without irritating it so that it does not trigger pain. Remedies with these properties are those that are fine of substance and not viscous, such as barley groats[22] (among foods) and filtered chicory juice[23] [*Cichorium endivia* and var. or *Cichorium intybus*] (among other remedies). But chicory syrup, although it is fine and not viscous, thickens the liver because of its sweetness, for all sweet things have the property that thickens the liver and spleen.[24] And apple syrup and pomegranate syrup, even though they are coolants, block the openings of the passages through their astringency and thus increase the obstruction.[25] A remedy does not reach remote organs unless it has been transformed and weakened in strength; this is not the case with organs that are close [to where the remedy enters the body]. The[26] eye, because of its extreme sensitivity, will tolerate only [medications] that alleviate pain, such as egg white and mucilage of fenugreek [*Trigonella foenum graecum*], and so also tolerates only those remedies that are extremely fat. Since this is the case, treatment necessarily differs according to the differences in organs and diseases.

(16) If someone's body contains a remnant of the material of a disease, it should be expelled; its [the body's] temperament should be balanced before reviving its strength. If it is suddenly emaciated through [excessive] evacuation, its strength can be revived in such a way that it regains its [initial] condition within one or two days, by replacing that which has departed from the body with a large quantity of suitable foods and agreeable smells. [This can happen because] the powers of organs of [such a body] have been preserved [intact], without any loss. If [someone's body] has become emaciated over a long period of time,[27] its [treatment] will take a long time [as well], because the powers of his organs are weak. Therefore, it is necessary to give him well-prepared bread along with chickens, fish, and the like, from among those [foods] that

(١٥) فصل: الوقوف على معرفة كلّ واحد من الأعضاء وما يخصّه نافع ضرورة في العلاج وذلك أنّ الكبد إذا ورمت ورما حارّا فهي تحتاج إلى ما يفتح ويجلو من غير تلذّع لئلا يهيج الوجع. والأدوية التي هي كذلك هي اللطيفة الجوهر التي لا لزوجة فيها إمّا من الأغذية فمثل كشك الشعير وإمّا من الأدوية فمثل ماء الهندباء المروّق
٥ وأمّا شراب الهندباء فإنّه وإن كان لطيفا لا لزوجة فيه فإنّه بحلاوته يغلّظ الكبد من قبل أنّ الأشياء الحلوة كلّها من شأنها أن تغلّظ الكبد والطحال. وأمّا شراب الرمّان والتفّاح فإنّهما وإن كانا يبرّدان فإنهما يضمّان فوّهة المجاري بالقبض الذي فيها فيزيد في الاحتقان وأيضا فإنّ الأعضاء البعيدة لا يصل إليها الدواء إلّا وقد استحال وضعفت قوته. وليس كذلك الأعضاء القريبة وأيضا فإنّ العين لفرط حسّها لا تحتمل
١٠ إلّا ما يسكّن الوجع كبياض البيض ولعاب الحلبة ولا تحتمل من الأدوية إلّا ما كان شديد السمن فإذا كان الأمر على هذا فواجب أن يكون العلاج يختلف باختلاف الأعضاء والعلل.

(١٦) فصل: إن كان البقية من قد بقا في بدنه بقية من مادّة المرض فينبغي أن ينفض وأن يعدّل مزاجه قبل الإنعاش فإن كان نهك ضربة من الاستفراغ فقد
١٥ يمكن أن تنعش قوته حتّى تصير إلى حالها في يوم أو يومين بأن يخالف عوضا (fol. 111b) ممّا ذهب من بدنه بالتوسّع في الأغذية والأرائح الموافقة وذلك أنّ أعضاءه محفوظة القوى لم ينقص منها شيء. وإن كان ينهك بدنه في زمان طويل فليس يمكن ذلك فيه إلا في زمان طويل لأنّ قوى أعضائه ضعيفة. ومن أجل هذا ينبغي أن يعطى الخبز المحكم الصنعة مع الفراريج والسمك ونحو ذلك ممّا هو سريع الانهضام

١ الوقوف] emend. eds. الاقوف MS || ١٣ البقية] emend. eds. الباقية MS

digest quickly, are absorbed quickly, and produce good humors. One should increase their quantity every day according to his capacity to bear [the increase in the amount of food]. One should also let him drink watery, astringent wine because it increases the total heat [of the body],[28] penetrates into the food, helps its digestion, and breaks up the winds. One should take care that the food does not weigh heavily on the stomach.[29] One should do all one can until [the patient's] condition is such that he can have the meat of a kid, after which comes the meat of a lamb.[30] He should be careful that the food does not float in the stomach as a result of drinking a large quantity of water.[31] If he suffers from severe thirst, let him drink a small amount of water mixed with wine.

(17) Arriving at the best treatment will come from [knowing] several factors, including:

i. The kind of disease [someone is suffering from], its cause, and [its] symptoms.

ii. The temperament [of an ill person]: If a hot temperament alters in the direction of heat [i.e., becomes even hotter], then it will return [to its natural state] through a small amount of cooling; but if a hot temperament alters in the direction of cold, then it will return only through a large amount of heating.

iii. The activities of the organs: The powers of bodily parts with noble activities should be preserved, and one should beware of evacuating [these parts] forcefully. For this reason, if we want to dissolve and evacuate [a residue from these parts], we mix the remedies [we use] with some aromatic, astringent ingredient that preserves the power of the bodily part, even if it affects the treatment adversely.

iv. The disposition [or nature] of the organs: The stomach is cleansed through both excretion and emesis; the rectum can be cleansed only by excretion, and the liver can be cleansed from its residues by excretion and urination.

v. The location of the parts: External [parts] can be reached by medicines before their powers are changed, whereas for most of the internal parts, the opposite is the case.

vi. The reciprocity of the parts: When the head hurts, the stomach shares [in the pain];[32] therefore, both the stomach and the head should be cleansed so that [the head] is not affected by the matter rushing to it from the stomach.[33]

والنفوذ مولّد للخلط المحمود ويزاد في ذلك كلّ يوم بمقدار احتمال قوته. وكذلك أيضا يسقى من الشراب المائي القابض لأجمع الزيادة في حرارته والتنفيذ في الغذاء والإعانة على الهضم والكسر للرياح. وينبغي أن تعتنى أن لا يثقل الطعام على معدته ويفعل حتّى تبلغه إلى لحم الجدي ومن بعده لحم الحمل. واحرز أن يطفوا الطعام في معدته بكثرة ٥ شرب الماء فإن اشتدّ بهم العطش فاسقهم من الماء المقدار اليسير ممزوجا بالشراب.

(١٧) فصل: أفضل اكتساب نوع العلاج يكون من أشياء منها: (ا) من نوع المرض وسببه وأعراضه (ب) من مزاجه وذلك أنّ المزاج الحارّ إذا تغيّر إلى الحرارة فردّه يكون بالمقدار القليل من التبريد والمزاج الحارّ إذا تغيّر إلى البرودة فردّه يكون بالمقدار الكثير من التسخين (ج) من أفعال الأعضاء فإنّ الأعضاء التي لها أفعال ١٠ شريفة ينبغي أن تحفظ عليها قواها ويحرز أن تستفرغ بقوة ومن أجل هذا إذا أردنا أن نحلّل منها شيئا ونستفرغه نخلط في الأدوية ما يحفظ قوة العضو من الأشياء العطرية القابضة وإن ضادّ ذلك بالعلاج (د) من خلق الأعضاء وذلك أنّ المعدة تنقّى بالبراز والقيء والأمعاء المنتصبة لا يمكن أن تنقّى إلا بالبراز وحده والكبد يمكن أن تنقّى فضولها بالبراز والبول (ه) من موضع الأعضاء فإنّ الظاهرة منها تلقاها ١٥ الأدوية قبل أن تتغيّر قواها والكثير من الأعضاء الباطنة خلاف ذلك (و) من مشاركة الأعضاء إذا ألم الرأس تشاركه المعدة فينبغي أن (fol. 112a) تنقّى المعدة وأن ينقّى الرأس حتّى لا يقبل ما تبادر إليه من المعدة (ز) من قوى الأعضاء فإنّ

٣ يثقل] .emend. eds يقثل MS || ٧ مزاجه] .emend. eds امزاجه MS || ٩ الأعضاء] .emend. eds الأفعال MS || ١٠ أن] ditt. MS | تستفرغ] .emend. eds تستفراغ MS || ١١ ونستفرغه] .emend. eds ونستفراغه MS

vii. The capacities of the parts: Parts affected by high fever cannot endure hard treatment, nor can weak parts.

viii. The remedies themselves: The dose of strong remedies should be calculated [precisely] and should be mixed with other ingredients according to the need, and so also for other remedies.

ix. The things that, through their conformity or difference, indicate what [sort of treatment] is needed.[34]

(18) One should evaluate the cause [of the ailment], the ailment [itself], and its symptoms, as each one stands in relation to the others; then one should pay attention to the [thing] that causes the strongest disturbance [in the body] and weakens its strength most of all. If you come across something that counteracts all these things together or that counteracts that which is most severe but is beneficial for the other things, you should rely on it. But if you do not come across something like that, you [should turn to treating] that which is most important without neglecting the other [things]. If the cause of the fever is alarming [very dangerous], one should take care to root it out, even if this [action] increases the heat of the fever. And sometimes the fever itself is so high and severe that one should hasten to cool it and extinguish it, even if this [treatment] worsens the cause. And in some cases the fever is accompanied by severe symptoms—such as a collapse of strength, excessive diarrhea, and fainting—and one should quickly attend to the symptom [first], so that it disappears, and then start treating the fever.

(19) If a humor flows into the stomach, it should be expelled from it using emesis or purgation. The influx should then be stopped through the evacuation of that humor from the body, and its cause should be neutralized. The stomach should be strengthened so that it does not absorb [the humor] anymore. If [the stomach] has been affected by a bad temperament because of the lengthy influx [of the humor], that temperament should be changed [into a better one] once [the stomach] has been cleansed. If the humor has entered into the lining [of the stomach] and adheres to it, [the condition] should be treated with

الأعضاء الشديدة الحمّى لا تحتمل العلاج الصعب وكذلك الأعضاء الضعيفة (ح)
من الأدوية أنفسها فإنّ الأدوية القوية ينبغي أن تقدّر كميتها وأن تخلط بها أشياء
أخر بحسب الحاجة وكذلك من سائر أنواع الأدوية (ط) من الأشياء التي باتّفاقها
واختلافها يستدلّ على ما يحتاج إليه.

٥ (١٨) فصل: ينبغي أن تقدّر السبب والمرض والعرض بعضها ببعض ويقصد إلى
أشدّها إزعاجا وأنهكها للقوة فإن صادفت ما يبطل جميعها معا أو يبطل أشدّها
وينفع الآخر فاعتمد عليه. وإن لم تصادف ذلك فإلى الأهمّ من غير أن يغفل الآخر
فإنّ الحمّى إذا كان سببها مخوّفا ينبغي أن تصرف العناية إلى قلعه وإن كان ذلك يزيد في
حرارة الحمّى. وأيضا فربّما كان بالحمّى نفسها من العظم أو الشدّة ما يفزع إلى التطفية
١٠ والتبريد وإن كان ذلك زائد في سببها. وكذلك ربّما كان مع الحمّى عرض خطر كسقوط
القوة وإفراط الذرب والغشي يهوج إلى الإقبال على العرض حتّى يزول ثمّ يؤخذ
في علاج الحمّى.

(١٩) فصل: إذا كان ينصبّ إلى المعدة خلط ينبغي أن يخرج منها إمّا بالقيء وإمّا
بالإسهال ثمّ يقطع ذلك الانصباب. وانقطاعه يكون باستفراغ ذلك الخلط من
١٥ البدن ويمنع الدافع وتقوّى المعدة حتّى لا تقبله فإن كان قد حدث فيها سوء مزاج من
طول انصبابه أن يتبدّل ذلك المزاج بعد نقائها وإن كان الخلط قد دخل جرمها ولج

---

٥ والمرض] emend. eds. والمراض MS || ٦ وأنهكها] emend. eds. وأهتكها MS | صادفت] emend. eds. صدفت MS || ٩ يفزع] emend. eds. يحرع؟ MS || ١٥ تقبله] emend. eds. تقبيله MS | حدث] emend. eds. حدوث MS

drastic[35] remedies, such as hiera[36] picra.[37] When the humor is viscous, it should be refined and diluted before its evacuation. Those [suffering from it] should be given such foods that cleanse the humor. One should proceed in this way with every single organ.

(20) The issue of food [to be given to someone ill] has six aspects:

i. One of these is the strength [of the patient]: If it is sound, then he can tolerate waiting and a lightening [of his diet]. The opposite [also holds true—if his strength is weak—then he must be given food right away, and his diet should not be lightened].

ii. The length of the disease: If the disease lasts for a long time, the [patient's] strength should be maintained by means of food.

iii. The disease [itself]: [The quality of the food given] should be opposite to [that of] the disease. For instance, in the case of fever caused by the putrefaction of thick, viscous humor, the food should thin out the thickness of the humor.

iv. The time: [The food] should be administered at the usual time, both during healthy [periods] and when the attack abates.[38]

v. The digestive organs: If the stomach or liver is affected by a tumor and food is administered before a [fever] attack, [the nourishment] is detrimental for the patient, especially when the body is overfilled. When one of [these organs] is weak because of a bad temperament or because of the influx of humor but is not affected by a tumor, food is appropriate, even during a [fever] attack, especially when the body is not congested.

vi. The magnitude of overfilling: When the body is overfilled, one should take less food; but when it is deficient, one should take more food, even during a fever attack.[39]

(21) The [superfluous] matter in the lower parts of the body is drawn upward and evacuated by exercising the upper parts that are opposite [to the lower parts]. The reverse holds true for the [superfluous] matter in the upper parts: [it is drawn downward by exercising the lower parts]. Cupping glasses also have a strong attractive force, upon their application. In keeping with this [principle],[40] they should be applied to the opposite side if the goal is to divert the matter [to that other side]. But if the goal is to draw the [superfluous] matter from the [afflicted] part [itself], they should be applied [on that part] either by using fire or with or without scarification.

فيها فينبغي أن يعطا الأدوية العواصة(!) كالإيارج الفيقرا وإن كان الخلط لزجا فينبغي
قبل استفراغه أن يلطّف ويقطع وينبغي أن تجعل أغذية هؤلاء فيها (fol. 112b) جلاء
لذلك الخلط. وعلى هذا المثال ينبغي أن يفعل في كلّ واحد من الأعضاء(؟).

(٢٠) فصل: أمر الغذاء يكتسب (و) وجوه: (ا) منها القوة فإنها إذا كانت صحيحة
٥ احتملت الصبر والتلطيف وبالضدّ (ب) مدّة المرض فإن كان الزمان فيه طويلا
وجب أن تحفظ القوة بالغذاء (ج) المرض فإنّه ينبغي أن يكون مداضّا له مثل الحمّى
الحادثة عن عفونة خلط غليظ لزج ينبغي أن يكون ملطّفا لغلظ الخلط (د) الوقت
فإنّه ينبغي أن يعطى في وقت العادة التي كانت في الصحّة وفي وقت فتور النوبة (ه)
من آلات الغذاء وذلك أنّ المعدة والكبد إذا كان أحدهما وارما كان الغذاء قبل
١٠ النوبة متلفا للمريض وخاصّة إذا كان البدن ممتلئا. وإن كان أحدهما ضعيفا من
سوء مزاج أو من انصباب خلط من غير ورم فالغذاء موافق في ذلك ولو في وقت
النوبة وخاصّة إن كان البدن ناقص الامتلاء (و) من مقدار الامتلاء فإنّ البدن
الممتلئ فينبغي أن ينقص من الغذاء والبدن الناقص ينبغي أن يزاد فيه الغذاء ولو في
وقت نوبة الحمّى.

١٥ (٢١) فصل: المادّة التي في الأعضاء السفلى تجذب إلى فوق برياضة الأعضاء
العليا على المحاذاة واستفراغها وبالضدّ في المادّة التي في الأعضاء العليا. والمحاجم
أيضا تجذب اجتذابا قويا إذا علقت فمن أجل ذلك ينبغي أن تعلق في الجهة المقابلة إذا
كان الغرض إمالة المادّة فأمّا إن كان الغرض جذب المادّة من العضو فينبغي أن تعلق
عليه إمّا بشرط وإمّا بلا شرط وإمّا بنار.

---

١ فينبغي] قبل استفراغه أن يلطف ويقطع وينبغي أن يقطع يجعل added and deleted MS | أن] ثمّ add. MS || ٥ فإن] إن add. MS || ٦ يكون مداضّا] emend. eds تكون مداضة MS || ١١ ورم] emend. eds. ورام MS

(22) Overfilling should be treated by bleeding, sweating, exercise, massage, diet, and remedies that evacuate [the residues] or that refine [the humors]. Bad humors should be treated by evacuation and [then by bringing the humors back] into equilibrium.

(23) If the body contains a very raw[41] humor, do not even consider[42] bleeding, lest the innate heat become too weak to concoct [the humor]. Likewise, do not think of exercise, purgation, or bathing, lest the fine part of the humor be expelled and the crude part remain and flow to one [or more] of the noble organs. One should remove [the humor] only by means of massage, which concocts [the humor] and which should be applied as follows: Start with the legs and rub them from the joint of the knee to the feet; then rub the thighs from the groin to the knee. Once these parts have been softened through rubbing and the body has become warm, rub the forearms, then the upper arms, followed by the spinal column. Then go back to the legs. [Repeat this] for the whole day, except for the time in which the patient takes a nap, for sleep is beneficial in this case because it concocts [the crude humor]. Accordingly, at the moment when the patient is fatigued, let him be rubbed with an oil that has a dissolving effect; then rub this off and let him rest. Once the [patient] has recovered from his fatigue, one should massage him again. Bodies of a moderate [temperament] should be massaged with linen cloths of moderate softness or roughness, and those of a dry [temperament] with rough linen cloths. Those with this bodily condition should live in a house that is only moderately hot and cold, for heat dissolves the humors, and one cannot be free from the fear that they will suddenly stream to a noble organ; and cold increases the [thickness] of the humors. In cases such as these, one should limit oneself to [the ingestion of] hydromel.[43] If [the patient also suffers from] fever, one

(٢٢) فصل: علاج الإمتلاء يكون بالفصد وبالتعريق وبالرياضة وبالدلك وبتقليل الغذاء والأدوية التي تستفرغ والتي تلطّف وعلاج (fol. 113a) رداءة الأخلاط يكون بالاستفراغ والتعديل.

(٢٣) فصل: إذا كان في البدن خلط فجّ كثيرا فلا تقرب الفصد لئلا تضعف
٥ الحرارة الغريزية إنضاجه ولا تقرب الرياضة والإسهال والاستحمام لئلا يخرج رقيق الخلط ويبقى غليظه وينصبّ إلى بعض الأعضاء الشريفة ولا ينبغي أن تنفضه ‹إلا› بالدلك الذي ينضج وهذا الدلك يكون على هذا المثال: ابدأ من الساقين وادلكهما من مفصل الركبة إلى القدم ثمّ ادلك الفخذين من الأربية إلى الركبة فإذا تسحّقت هذا الأعضاء وسخن البدن فادلك الساعدين ثمّ العضدين ثمّ الصلب ثمّ عد إلى الساقين
١٠ وافعل ذلك نهارا أجمع إلّا في الوقت الذي يأخذ فيه العليل النوم فإنّ النوم نافع في هذا الحال لأنّه ينضج وكذلك الأوقات الذي يأخذ العليل الإعياء فأمر فيه بالدهن الذي يحلّل ثمّ يمسح ويراح. فإذا سكن الإعياء فينبغي أن يعاد إلى الدلك. ويكون الدلك في الأبدان المعتدلة بالمناديل متوسّطة في اللين والخشونة وفي الأبدان اليابسة بمناديل خشنة وأصحاب هذا الحال ينبغي أن تكون مساكنهم معتدلة في
١٥ الحارّ والبرد وذلك أنّ الحرارة تذيب الأخلاط فلا يؤمن انصبابها دفعة إلى عضو شريف والبرد يزيد في نهوة(!) الأخلاط. وتقتصر بهؤلاء على ماء العسل فإن كان حمّى يجعل بدل ماء العسل السكنجبين ويكون غذاؤهم حساء متّخذا من حنطة وإن

---

٤ البدن] .emend. eds البدان MS || ٧ الساقين] .emend. eds السقين MS || ٨ تسحّقت] .emend. eds اتسخفت MS || ٩ الساقين] .emend. eds السقين MS || ١٠ نهارا] .emend. eds نهري MS || ١١ يأخذ] .emend. eds يأخذه MS || ١٣ بالمناديل] .emend. eds بالمنادل MS

should administer oxymel instead of hydromel, and [these patients] should feed themselves with gruel prepared from wheat.[44] If they suffer from diarrhea, one should give them pomegranate seeds or sumac [*Rhus coriaria*] together with the gruel; if the [diarrhea] becomes severe, [one] should moisten [these ingredients] in astringent, diluted wine. In general, the treatment of these [patients] should consist of refining, dissolving, and evacuating the [crude] humors. [However], if one of them suffers from a tumor in the liver or stomach, there is no hope that he will be saved.

(24) Those whose humors are extremely thin and fluid and whose bodies are quick to dissolve and putrefy should take food in small amounts and repeatedly. Once the first [portion of] food has passed [through the body], they should take a second. If their organs are weak, they cannot take their food all at once; and if they suffer from a tumor in the stomach or liver, they cannot be saved. Their slenderness results from the wideness of their pores and the weakness of their vigor. It should therefore be counteracted with that which thickens the body and strengthens it, externally or internally, by taking astringent ingredients; the thinness of the humors should be counteracted with that which thickens them, and that which has dissolved from [the body] should be replaced through nutrition.

(25) If pain arises from a biting humor, its harm should be undone by those purgatives that remove, dissolve, and cleanse that humor. If this is not sufficient, [one should use] that which benumbs the senses [—namely, a narcotic]. If [the pain] comes from a thick, viscous humor, it should be alleviated through dissolving, concocting, and evacuating that humor with thinning remedies that do not heat excessively. If[45] [the pain] comes from inflating wind, it should be treated with foods, remedies, cataplasms, and fomentations that disperse the winds.

استطلقت بطونهم فاجعل لهم مع الحساء حبّ رمّان أو سمّاق فإن أفرط فينبغي أن ينبلّ في شراب قابض ممزوج وبالقول المطلق فإنّ علاج هؤلاء تلطيف الأخلاط وتقطيعها واستفراغها ومن كان من هؤلاء به ورم في الكبد أو في المعدة فلا تطمع له بخلاص.

٥ (٢٤) فصل: الذين أخلاطهم في (fol. 113b) غاية الرقّة والسيلان وأبدانهم سريعة التحليل والعفن يسرع إليهم ينبغي أن يغتذوا قليلا قليلا مرّة بعد أخرى كلّما استمرّ الغذاء الأوّل أخذ الغذاء الثاني. فإن ضعفت أعضاؤهم لا يمكن معه أخذ الغذاء في دفعة واحدة وهؤلاء أيضا إن كان في معدهم وأكبادهم ورم فلا خلاص لهم. والانخراط في هؤلاء يكون من سعة المسامّ وضعف القوة ولذلك يجب أن ١٠ تقابل هذا بما يكثّف البدن ويقوّيه إمّا من خارج وإمّا من داخل فيأخذ الأشياء القابضة وأن تقابل رقّة الأخلاط بما يغلّظه وأن يخلف عليهم بالتغذية عوض ما انحلّ منهم.

(٢٥) فصل: الوجع إن كان من خلط لذّاع فينبغي أن تقلع عاديته بما يقلع ذلك الخلط ويحلّله ويغسله من المسهلات فإن لم يف ذلك فبما يخدّر الحسّ. وإن كان من ١٥ خلط غليظ لزج فيسكّن بتقطيع ذلك الخلط وإنضاجه واستفراغه بالأدوية التي تلطّف من غير إسخان مفرط. وإن كان من ريح نافخة فيعالج بالأغذية والأدوية والأضمدة والنطولات التي تفشّ الرياح.

---

٢ ينبلّ] emend. eds. ينبيل MS || ٧ ضعفت] emend. eds. ضعف MS || ٩ سعة] emend. eds. ساعة MS || ١١ عوض] emend. eds. عرض MS || ١٣ تقلع] تكتسر MS[2]

(26) One should protect against fainting (syncope) by preserving the substance of the pneuma,[46] the faculties of the organs, and the innate heat. This can be achieved with foods and sweet, pleasant smells; thickening the body when it is porous and thickening the humors when they are thin; thinning the humors when they are thick and thinning the body when its pores are tight; alleviating pain when it is severe; and, in general, by preventing that which causes fainting.

(27) If [someone's] stomach has a hot dyscrasia, he should be given cold water to drink[47] and fed with cooled sour milk, fruit that has been cooled in the snow or in the cold air, and cooled barley groats. A poultice containing ingredients of this sort should be applied to his stomach. But if it is a cold dyscrasia, he should be given old, pure wine to drink; he should be fed and given a poultice, [both the food and poultice] being of moderately hot ingredients; and his belly should be brought into contact with [things] that heat it moderately until his body becomes well fleshed. If the dyscrasia is moist, he should be given roasted food, and a poultice of drying ingredients should be applied [to his stomach]; but if the dyscrasia is dry, one should give him moist things. If the dyscrasia is composite, the treatment should be composite as well.[48]

(28) Fever is a hot, dry dyscrasia, and it requires [for its cure] cold, moist things, unless [such treatment] would lead to a thick, viscous humor or weakness of a noble organ that cannot bear cooling. It may also result in an affliction worse than the fever, such as fainting. Therefore, that which is more important should have precedence, but the other [things] should not be neglected.

(٢٦) فصل: ينبغي أن يتحفّظ من الوقوع في الغشي بحفظ جوهر الروح وقوى الأعضاء والحرارة الغريزية وذلك يكون بالأغذية والأرائح اللذيذة وبتكثيف البدن وتغليظ الأخلاط إذا كانت رقيقة والبدن متخلخلا وبتلطيف الأخلاط وبتسخيف البدن إذا كانت غليظة والبدن مستحصف المسامّ وبتسكين الأوجاع إذا كانت ٥ مفرطة وبالجملة أن يمنع السبب الفاعل للغشي.

(٢٧) فصل: إذا كان في معدة سوء مزاج حارّ فينبغي أن يسقى صاحبها الماء البارد وأن يطعم اللبن المحيض المبرّد والفاكهة التي قد برّدت على الثلج أو في الهواء البارد وكشك الشعير المبرّد وأن تضمّد المعدة من خارج بمثل ذلك. وإن كان فيها سوء مزاج بارد فينبغي أن يسقى صاحبها الشراب العتيق الصرف وأن يغذّى (fol. 114a) ويضمّد ١٠ بالأشياء الحارّة وأن يضمّ إلى بطنه ما يسخنه باعتدال حتّى خصب البدن وإن كان فيها سوء مزاج رطب فينبغي أن يعطى القلايا وأن يضمّد بالأشياء المجفّفة وإذا كان فيها سوء مزاج يابس ينبغي أن يعطى الأشياء الرطبة. وإذا تركّب سوء المزاج ينبغي أن يركّب العلاج.

(٢٨) فصل: الحمّى سوء مزاج حارّ يابس فهو يحتاج الأشياء الباردة الرطبة ١٥ إلّا أنّه قد يعرض من ذلك إمّا خلط غليظ لزج وإمّا ضعف عضو شريف لا يحتمل التبريد وإمّا عرض آخر أشدّ من الحمّى كالغشي ومن أجل هذا ينبغي أن تقدّم الأهمّ من غير أن يغفل الآخر.

---

١٠ خصب] emend. eds. خصبت MS | البدن] emend. eds. البدان MS

(29) Regarding every fever, observe the change it makes in the temperament in view of the nature of the fever; then bring it [the temperament] into balance. In general, the cause of every fever should be opposed. For instance, although the cause is no longer active in most cases of ephemeral fever, it is still necessary to [oppose it]. If someone has fever because of sleeplessness, one should make him sleep; if someone has fever because of anxiety, one should give him pleasure. The other kinds of ephemeral fever should be opposed in the same way.

(30) Cold water is appropriate for fever patients because it extinguishes a flare-up by cooling and moistening. [It is also appropriate for] severe anxiety and distress. It strengthens the intestines and dissolves biles in them, and it invigorates the body from [the weakness caused by] heat and dryness and prepares it for the intake of food. But one should not administer [cold water] when the noble organs are weak because of a tumor or because of a cold or moist dyscrasia. [And], by God, certainly not [when] the tumor is [...], and not before [the tumor] is ripe, nor during the crisis, in order to allow it to ripen. But if the body is well fleshed, dry heat dominates, the humors are concocted and may [already] have been evacuated, and there is no longer any weak organ within; then cold water should be given to drink.

(31) Once the ephemeral fever has declined, it should be treated by [going to] the bathhouse in order to dissolve the smoky vapor that arises in the body.[49] If the fever is caused by bathing in astringent water, plenty of massage should be applied in order to relax the tightness [of the skin]. If [it is caused by] anger or anxiety, the massage should be moderate and the rubbing with oil should be extensive. When[50] [the fever] arises from cold, there is more need for hot air; but if it is accompanied by a catarrh or rheum and has not concocted, bathing should be postponed, and one should make the head sweat by [rubbing] it with hot oil. Someone who has developed a fever because of the heat of poisons should be immersed

(٢٩) فصل: إقصد في كلّ حمّى إلى تبديلها المزاج من حيث طبيعة الحمّى ثمّ تعديله. وبالجملة فينبغي أن تضادّ سبب كلّ حمّى فإنّ حمّى يوم التي سببها قد بطل على الأكثر يحتاج إلى مثل هذا وذلك أنّ من حمّ من سهر ينبغي أن يجلب له النوم ومن حمّ من همّ ينبغي أن يجلب له السرور وكذلك في سائر أصناف حمّى يوم ٥ ينبغي أن تضادّ.

(٣٠) فصل: الماء البارد يوافق أصحاب الحمّيات لأنّه يطفئ الالتهاب ببرده وترطيبه وكثرة القلق والكرب ويقوّي الأحشاء ويذيب ما فيها من أمرار وينعش البدن من الحرارة واليبس ويعدّ البدن للاغتذاء إلّا أنّه لا ينبغي أن يسقى إذا كانت الأعضاء الشريفة ضعيفة من ورم أو من سوء مزاج بارد أو رطب اللهمّ ألا يكون ١٠ الورم ‹. . .› ولا قبل النضج ولا في وقت البحران لئلا يعوق عن النضج. فأمّا إذا كان البدن خصيبا وقد استولت الحرارة ويبسها وأخلاطه بنضج وقد استفرغت وليس فيه عضو ضعيف فينبغي أن يسقى الماء البارد.

(٣١) فصل: حمّى يوم يحتاج في علاجها إلى الحمّام بعد انحطاط الحمّى ليتحلّل ما يتولّد (fol. 114b) في البدن من البخار الدخاني فإن كان سبب الحمّى الاستحمام بالمياه ١٥ القابضة ينبغي أن يكون الدلك كثيرا بما يزيل الاستحصاف وإن غضبا أو همّا ينبغي أن يكون الدلك متوسّطا وأن يكون المرخ بالدهن كثيرا وإن كان من برد فالحاجة إلى الهواء الحارّ أكثر فإن كان معها نزلة أو زكام ولم تنضج فينبغي أن تأخّر الحمّام حتّى تنضج النزلة وأن يعرّق الرأس بدهن مسخّن. ومن حمي من وهج السموم ينبغي أن يغمس في

٩ ألا] emend. eds لا MS || ١١ وأخلاطه] emend. eds. وأخلاطها MS || ١٥ أن] أن ditt. MS || ١٦ من برد] emend. eds. مبردا MS

in cold water and eat cold fruits before going into the bathing basin and after leaving it. If the fever arises from the frequent getting up because of diarrhea, [the patient] should be fed before entering the bathhouse. One should pay attention to the stomach, especially if it is affected by a burning [sensation].[51] The same holds good for the intestines, especially if they are irritated. If [the fever] comes from indigestion, one should inquire whether the food has degenerated into something smoky or sour. Each [quality—smoky or sour]—should be countered with its opposite. In any case, one should take care that the putrefied food is eliminated before [the patient] enters the bathhouse; after that, the stomach should be strengthened. If the fever comes from a tumor in soft flesh, that tumor should be dissolved before [the patient] enters the bathhouse.

(32) Those indigestions that are beneath the ribs should be confined. If the food has descended from the stomach to the intestines, one should give the patient with this condition an enema whose ingredients expel. But if it has not gone down, one should give him that which induces emesis and let him drink a pepper stomachic and the like. One should warm his stomach and apply a strengthening poultice to it. If [the food] is concocted in it, the stomach should be rubbed with olive oil in which rue or seeds that expel the winds were cooked.[52] When the food has descended and left the body, [the patient] should have some fine [food] immediately and then, about six hours later, enter the bathhouse. Know that if the treatment of ephemeral fever is not successful, it will either last for days or lead to putrefaction.

(33) When a fever occurs, one should examine it. If it is an ephemeral fever, the patient should enter the bathhouse when the fever abates, [according to] the opinion of all physicians. But I caution against taking him to the bathhouse, because we know little about this nowadays. And although there may be someone who knows about the nature of fevers and about the effect of [going to the bathhouse], we do not know about it

الماء البارد وأن يأكل من الفواكه الباردة قبل دخول الأبزن وبعد الخروج منه. وإن كان الحمّى من كثرة التردّد في القيام في الذرب ينبغي أن يغذى قبل دخول الحمّام وأن يعنى بمعدهم وخاصّة إن كان فيها لذع وكذلك بالأمعاء ولا سيّما ‹ما› قد أجحف. وإن كان من تخمة ينبغي أن ينظر هل يفسد الطعام إلى الدخانية أو إلى الحموضة فتقابل كلّ واحدة منها بضدّه وتنظر على كلّ حال إخراج الطعام الفاسد قبل دخول الحمّام ثمّ تقوّى المعدة. وإن كان الحمّى من قبل الورم في اللحم الرخو ينبغي أن تحلّل ذلك الورم قبل دخول الحمّام.

(٣٢) فصل: التخم ينبغي أن تحبس منهم ما تحت الشراسيف فإن كان الطعام قد انحدر عن المعدة إلى الأمعاء ينبغي أن يحقن صاحب هذا الحال بما يخرجه وإن كان لم ينحدر فينبغي أن يعطى ما يجلب القيء وأن يسقى جوارش الفلافلي ونحوه وأن تسخّن المعدة وتضمّد بما يقوّيها وإن كان فيها نضج ينبغي أن يمرخ البطن بزيت قد طبخ فيه سذاب وبزور طاردة للرياح. فإذا انحدر الطعام (fol. 115a) وخرج عن البدن فينبغي أن يغذى البدن على المكان بشيء لطيف ثمّ يدخل الحمّام بعد ذلك بقدر وساعات. واعلم أنّ حمّى يوم إن أخطئ في علاجها إمّا أن تطول أيّاما وإمّا أن تؤول إلى العفن.

(٣٣) فصل: إذا أحدثت حمّى ينبغي أن يتفقّد أمرها فإن كانت حمّى يوم ينبغي أن يدخل صاحبها الحمّام وقت انحطاط الحمّى. هذا رأي جميع الأطبّاء وأمّا أنا فأحذّر من إدخالهم الحمّام لقلّة معرفتنا في زماننا هذا فإن كان عارفا بطبع الحمّيات وعارفا

٢ الحمّى] emend. eds. الحمر MS || ٣ وكذلك بالأمعاء] ditt. MS || ١٢ وخرج] emend. eds. وخراج MS || ١٧ رأي] emend. eds. وافى MS || ١٨ من] emend. eds. أن MS

[even] if we look into it, let alone knowing the effect. If [the patient suffers from] blood [fever, that is, synochous fever], one should hasten to bleed him.[53] If [the fever originates from] yellow bile, the humor from which it originates should be evacuated using ingredients that have the property to cleanse and evacuate without heating, such as tamarind [*Tamarindus indica*], pears, pomegranate juice, oxymel, barley groats, and spinach. If [the fever] is chronic, one should either, in the beginning, give the patient ingredients that refine the coarseness of the humor [that causes that fever], thin its viscosity, and help its concoction; or, in the end, when the humor is concocted, give him ingredients that evacuate it. If [the patient suffers from] hectic fever, one should take care to cool and moisten [his body] and to revive his vigor.

(34) Those who suffer from hectic fever benefit from [going to] the bathhouse because it clears their bodies of smoky vapor and moistens their limbs. If one wants them to enter the bathhouse, one should sprinkle much sweet water that is moderately hot in it so that the air is saturated with moist, sweet vapor. One should make the second room hotter than the first, and the third hotter than the second, and the fourth hotter than the third; but none should be excessively hot. The patient should be given accommodations very close to the bathhouse. Once that has been done, put him on a piece of cloth and tell him to undress in the middle room. Pour lukewarm oil on him and rub him gently with it. While he is [lying] on a stretcher, put him into moderately hot water by lowering the stretcher into the bathing basin so that it is immersed [in the water]; then lift him up and immerse him again, and do this repeatedly until he is close to sweating. When he is close to sweating, immerse

لما يؤول وأمّا نحن فليس نعرفها إذا عاينّاها فضلا أن نعرف لما يؤول. وإن كانت
دموية ينبغي أن يبادر صاحبها بفصد وإن كانت صفراوية ينبغي أن يستفرغ الخلط
الذي تولّدت منه بما من شأنه أن يجلو ويستفرغ من غير إسخان مثل التمر الهندي
والإجّاص وماء الرمّان والسكنجبين وكشك الشعير والإسفاناخ وإن كانت مزمنة
٥ ينبغي أن يعطى إمّا في أوّل الأمر ما يلطّف غلظ ذلك الخلط ويقطع لزوجته ويعين
على أنضاجه وإمّا في آخر الأمر عندما ينضج الخلط ما يستفرغه وإن كانت حمّى دقّ
فينبغي أن يعنى بالتبريد وترطيب وإنعاش القوة.

(٣٤) فصل: أصحاب حمّى الدقّ ينتفعون بالحمّام لأنّه يحلّل من أبدانهم البخار
الدخاني ويرطّب أعضاءهم. وإن أردت أن تدخلهم الحمّام فينبغي أن ترشّ الحمّام
١٠ بالماء العذب المعتدل الحرارة رشّا كثيرا حتّى يمتلئ الهواء من البخار الرطب العذب
واجعل البيت الثاني أسخن من الأوّل والثالث أسخن من الثاني والرابع أسخن من
الثالث ولا يكون واحد منهم مفرط الإسخان وأن تجعل تنزيل المريض بالقرب من
الحمّام جدّا فإذا فعلت ذلك فاجعل المريض على مقرمة وأمر أن يعرّى المريض في البيت
الأوسط وأن ينصبّ عليه دهن مفترّ ويمرخ فيه مرخا رقيقا ثمّ أنزله وهو على المحفّة
١٥ في الماء الحارّ المعتدل الحرارة بأن ترخي المحفّة في الأبزن حتّى تنغمس فيها ثمّ ارفعه
واغمسه أيضا وافعل ذلك أيضا مرّة إلى أن يقارب أن يعرق فإذا قرب من العرق

٣ | الذي] emend. eds. التي MS | ويستفرغ] emend. eds. ويستفراغ MS || ٦ ما] emend. eds. فيما MS || ١٢ منهم] emend. eds. منهما MS || ١٣ مقرمة] emend. eds. مضرامة MS || ١٦ واغمسه] emend. eds. واغمضه MS

him in cold water all at once, lift him up quickly, and rub him off immediately with moist towels so that the water goes away; then dress him and carry him on the stretcher to his accommodations. If he follows this regimen, his limbs will acquire much moisture without any harm.

(35) Someone whose body is very emaciated or dominated by a dry dyscrasia should be given milk to drink. The best milk is that of women—after that, the milk of a donkey, and then the milk of a goat. The younger the animal, the better the milk, especially if it has been well fed and well provided for. The best milk to use is that which is sucked from the breast; and if that is impossible, one should bring the animal close to the patient so that he can [drink] the milk the moment it is milked, while it is still hot and has not cooled off.[54] If one is afraid that it will curdle in his stomach, give [the patient] something with it that prevents [curdling] and that causes [the milk] to be quickly absorbed into the organs, such as honey and fresh milk. The purer the honey, the better it is, but sugar is [even] better.[55] [The milk] should be administered as follows: The patient should drink it, then rest for about four hours, and then go into the bathing basin once the milk has been digested and has gone down from the stomach. And when he has left the bathing basin, he should drink [milk] for a second time; but if he loathes it [this time], then leave him alone [do not force him to drink] and replace it with barley gruel and the like.

(36) When the strength [of the body] supports it, bleeding is beneficial in two ways for all fevers caused by repletion: (i) the evacuation of the [putrefied] matter and (ii) the extinction of the heat of the fever. The putrid humor should be expelled from the body through purgation, emesis, or micturition. What remains of it should be improved through nutrition and medication.

اغمسه في الماء البارد دفعة واحدة وارفعه بسرعة وامسحه من ساعته بمناديل رطبة
حتّي يذهب عنه ذلك الماء ثمّ لبّسه ثيابه واحمله على المحفّة إلى منزله وإذا دبّر بهذا
التدبير تكتسب أعضاؤه رطوبة كثيرة من غير ضرر.

(٣٥) فصل: إنّما ينبغي أن يسقى اللبن من قد نهك بدنه جدّا ومن قد غلب عليه
٥ سوء المزاج اليابس. وأجود الألبان لبن النساء ومن بعده لبن الأتن ومن بعده لبن
الماعز وكلّ من كان من هذا الحيوانات في سنّ الشباب كان لبنه أجود وخاصّة
إن أصلحت أغذيته وأحسن القيام عليها. وأبلغ ما يستعمل اللبن المصّ من الثديين
فإن لم يمكن فينبغي أن يقرب الحيوان من المريض ليأخذ اللبن ساعة يحلب وهو بعد
حارّ لم يبرد. فإذا خفت أن يتجبّن في المعدة فاسق معه ما يمنع من ذلك وما يسرّع
١٠ نفوذه ويدرّقه إلى الأعضاء مثل العسل والحليب وكلّما كان العسل أصفى كان أجود
والسكر أحسن منه. وصفة سقيه أن يشربه العليل ويستريح نحو أربع ساعات ثمّ يدخل
الأبزن في وقت استمرائه اللبن وانحداره عن المعدة فإذا خرج من الأبزن فيشربه
ثانية فإن عافته نفسه فينبغي أن يتركه ويعوّض في المرّة الثانية بمثل ماء الشعير ونحوه.

(٣٦) فصل: الفصد نافع في جميع الحمّيات الامتلائية متى ساعدت القوة من
١٥ وجهين: أحدهما استفراغ المادّة والثاني إطفاء حرارة الحمّى والخلط العفن ينبغي أن
يخرج ما يتعفّن منه عن البدن إمّا بالإسهال وإمّا بالقيء وإمّا بالبول وتصلح ما بقي منه
بالأغذية (fol. 116a) والأدوية.

---

٢ على] emend. eds. إلى MS || ٦ أجود] emend. eds. أجواد MS

(37) When fevers originate from a surplus of blood, one should apply venesection until the patient is on the verge of fainting, if his strength allows it.[56] When fevers originate from a humor mixed with blood, bleeding is also beneficial. For this reason bleeding is beneficial for all continuous fevers in which the humors are concocted.

(38) When the body of the patient is weak, when he suffers from indigestion, or when his body contains a crude humor that has not concocted, one should not bleed [that patient]. In the case of indigestion, one should postpone bleeding until the corruption of the food has diminished. Crude, viscous humors do not respond to expulsion through bleeding; and when one bleeds a body with weak organs, it becomes even weaker.[57]

(39) If there is a surplus of blood in the body, it should be evacuated all at once, if possible. But if this is not possible, [it should be done] in several steps. If [the blood] is bad in quality, it should be evacuated little by little. Each [quantity] that is evacuated from the body will be replaced with wholesome blood.[58]

(40) In all kinds of treatment, one should pay attention to the condition of the air, the external condition of the body, and the age [of the patient]. Thinness and a hot temperament dissolve the body; and a thin body that is affected by much dissolution, such as the body of a child, does not endure bleeding. The same holds good for a body with weak organs, such as the body of an old man, especially since an old man has a cold temperament and a small quantity of blood. If bile is produced in the cardia of the stomach of a body, [the person] is quick to vomit [and to develop] headache and syncope.[59]

(41) If you encounter other things that require bleeding, even after the seventh [day], do not pay attention to the number of days that have passed. Instead, you should determine the quantity to be bled, [keeping in mind] the other types of therapy, according to the strength [of the

(٣٧) فصل: ما كان من الحمّيات حادثا عن كثرة الدم فينبغي أن يخرج الدم إلى
أن يبلغ المريض الغشي إذا كانت القوة تحتمله وما كان من الحمّيات حادثا عن خلط
مخالط للدم فالفصد أيضا نافع. ومن أجل هذا صار الفصد في الحمّيات الدائمة كلّها
التي أخلاطها نضيجة نافعا.

٥ (٣٨) فصل: إذا كان بدن العليل ضعيفا أو كان به تخمة أو كان في بدنه خلط غليظ
لم ينضج فلا ينبغي أن يفصد وذلك أنّ التخمة ينبغي أن يؤخّر فيها الفصد إلى أن يؤول فساد
الطعام. والأخلاط الغليظة اللزجة لا تجيب إلى الخروج بالفصد والبدن الضعيف
الأعضاء إذا خرّج منه الدم زاد في ضعفه.

(٣٩) فصل: الدم إذا كان زائد الكمية فينبغي أن يستفرغ المقدار الزائد منه في
١٠ دفعة واحدة إن أمكن فإن لم يمكن في دفعات. وإن كان رديء الكيفية يستفرغ
قليلا قليلا وكلّما أستفرغ منه شيء أخلف على البدن عوضه دم جيّد.

(٤٠) فصل: يتفقّد في جميع ضروب العلاج حال الهواء وسحنة البدن والسنّ
فإنّ الرقّة والمزاج الحارّ يحلّ البدن والبدن النحيف الكثير التحلّل كبدن الصبي لا يحتمل
الفصد وكذلك البدن الضعيف الأعضاء كبدن الشيخ ولا سيّما أنّ الشيخ بارد المزاج
١٥ قليل الدم والبدن الذي يتولّد في معدته مرار فإنّه سريع القيء والصداع والغشي.

(٤١) فصل: إذا صادفت إلى سائر الأشياء تجب إلى الفصد ولو بعد السابع
ولا تلتفت إلى ما مضى من عدد الأيّام وإنّما ينبغي أن تقدّر الفصد لسائر أنواع

٩ يستفرغ] .emend. eds يستفراغ MS || ١٠ الكيفية] .emend. eds للكيفية MS | يستفرغ] .emend. eds يستفراغ MS || ١١ أستفرغ] .emend. eds أستفراغ MS || ١٣ الرقّة] .emend. eds الرقّة الحارّ MS || ١٦ تجب] .emend. eds تجيب MS

patient] and the quantity of the humor, in order to find out what should be done first and what later.

(42) If much blood has flowed to one of the organs and [the organ has] become occluded, the patient should be bled as soon as his strength and the other [conditions] are conducive.[60] If bleeding is impossible, [one should apply] cupping glasses. If the organ has become congested, one should employ solvents after attempting to evacuate it.

(43) If there is some hardness in one of the organs and it is dominated by heat, it should be softened with ingredients that are not that hot. If the hardness and rigidness are less thick, but the heat is very strong, the solvents should be mixed with remedies that have a discutient effect, such as vinegar. If one wants to soften something that lies within a noble organ, one should mix the solvent with strengthening remedies so that its strength is preserved.

(44) When one applies poultices with pure, astringent ingredients for the [treatment of] the internal organs, one should mix them with softening ingredients so that the strength of the external poultice reaches [those organs]. In the same way, remedies for [the treatment of] the remote organs should be mixed with remedies that make [them] reach [those organs] speedily.

(45) If there is a tumor in the stomach or liver accompanied by fever, it should be dissolved by means of solvent remedies that strengthen the organs. When the stream [of putrid matter] has stopped, the dissolving effect should be increased. If there is a tumor in the respiratory organs, the astringent effect of the poultices put on the chest should be less strong, so that the tumor does not rise to the heart or to the lungs. If there is a tumor in the lungs, one should beware of astringent ingredients in general, so that the [putrid] matter does not rise to the heart. One should mix the softening ingredients with ingredients that have a strong heating effect so that the [putrid] matter is drawn outward. [One should draw out the putrid matter] either through the application of cupping glasses or through other means. One should do so only once [the body] has been emptied so that [this kind of treatment] does not

العلاج بمقدار القوة وبمقدار الخلط ليعلم أيّ شيء يجب أن يقدّم وأيّ شيء يؤخّر.

(٤٢) فصل: إذا انصبّ واحتقن في بعض الأعضاء دم كثير (fol. 116b) فينبغي أن يفصد من ساعته العليل متى ساعدت قوته وسائر الأشياء فإن لم يمكن الفصد فالمحاجم فإن لج ذلك العضو فينبغي أن يستعمل فيه الأشياء المحلّلة بعد الاستفراغ.

(٤٣) فصل: إذا كان في أحد الأعضاء صلابة وقد غلبت عليها حرارة فينبغي أن تجعل التحليل بأشياء أقلّ حرارة وإن كان الجساء والصلابة أقلّ غلظ فاخلط بالأدوية التي تحلّل أدوية فيها تقطيع كالخلّ إذا كانت الحرارة أغلب. ومتى رمت تحليل ما في أحد الأعضاء الشريفة فاخلط في أدويتها أدوية مقوّية لتحفظ عليها قواها.

(٤٤) فصل: الأعضاء الباطنة تحتاج في أضمدتها بالأشياء القابضة الزكية إلى أن تخلط بها أشياء لطيفة لتوصل قوة الضماد من خارج إليها وكذلك ينبغي أن تخلط بأدوية الأعضاء البعيدة أدوية تدرّقها وتوصلها إليها.

(٤٥) فصل: إذا كان في المعدة والكبد ورم مع حمّى ينبغي أن يحلّل بأدوية فيها تحلّل وتقوية الأعضاء وإذا كان انقطع الانصباب ينبغي أن يزاد في التحليل وإذا كان في آلات التنفّس ورم ينبغي أن يجعل القابض في ضمادات الصدر أقلّ لئلا يرفع الورم إلى القلب أو إلى الرئة وإذا كان الورم في الرئة ينبغي أن يتوقّى الأشياء القابضة بالجملة لئلا يرفع المادّة إلى القلب وأن يخلط بالأشياء المرخية أشياء أشدّ إسخانا لتجذب المادّة إلى خارج إمّا بتعليق المحاجم وإمّا بغير ذلك وإنّما ينبغي أن يفعل

٨ تقطيع] emend. eds. تقطع MS || ١٢ وكذلك] emend. eds. وقذلك MS

attract [superfluous] matters to the chest, especially when the body is replete [with those matters].

(46) In the case of a hot tumor, the cooling limit [should be determined by] the magnitude of the heat of the tumor. If the cooling goes so far that the organ is on the verge of compacting, one should diminish [the degree of] cooling or stop it altogether so that it does not lead to hardness [of the tumor]. The limit of dissolving a hot tumor that has reached its climax is that the heating effect be of the magnitude of the [superfluous matter] evacuated from the [affected] organ. If the heating effect is such that it attracts other matter, one should diminish [the degree of heating]. One should turn to solvents only after the body has been emptied and the humors balanced.

(47) A hard tumor should be treated [first] by a softener, then by a discutient, and finally by a solvent. In general, one should see to it that the fine humors are not expelled and that the coarse humors do not remain, turning hard as stone and no longer reacting [to the solvents].[61]

(48) Scrofula affecting flesh that serves as filler (such as the flesh of the neck, the armpits, and the groin) should be treated by solvents alone, whereas scrofula affecting flesh that serves some useful purpose (such as the flesh of the breasts and [the flesh] that produces sputum) should be treated at first with repelling and strengthening [remedies], and then, when they reach their climax, with solvents. In general, all scrofula that do not react to solvents should be left to putrefy and then [should be] extirpated through surgery. During surgery one should be careful not to injure a nerve, artery, or vein.[62]

(49) Some atheromas contain a very thin and fine honeylike moisture; they should be treated with remedies that are powerful solvents. Others contain a very thick moisture; one should let these putrefy and then make an incision because solvents [alone] are not sufficient. The

ذلك بعد الاستفراغ كيلا تجتذب بهذا الأشياء الموادّ إلى جهة الصدر وخاصّة إن كان البدن ممتلئا.

(٤٦) فصل: الحدّ في التبريد في الورم الحارّ هو أن يكون بمقدار حرارة الورم وإن يبلغ التبريد إلى أن يميل العضو إلى الكمودة ينبغي أن ينقص منه أو يقطع لئلا يؤول الأمر (fol. 117a) إلى الصلابة. والحدّ في التحليل في انتهاء الورم الحارّ هو أن يكون التسخين بمقدار ما يستفرغ ما حصل في العضو فإن بلغ التسخين إلى اجتذاب مادّة أخرى ينبغي أن ينقص منه ولا يقرب شيئا من الأدوية المحلّلة إلّا بعد استفراغ البدن وتعديل الأخلاط.

(٤٧) فصل: ينبغي أن يعالج الورم الصلب بالتي تلين ثمّ بالتي تقطع ثمّ بالتي تحلل وبالجملة ينبغي أن يجعل الغرض فيه أن لا يخرج الرقيق من الخلط ويبقى الغليظ فيتحجّر ولا يجيب إلى التحليل.

(٤٨) فصل: الخنازير ما كان منها في لحم كالحشو مثل لحم العنق والإبط والأربية فعلاجه التحلّل فقط وأمّا ما كان منها في لحم له منفعة كلحم الثديين والمولّد للريق فعلاجه إمّا في الابتداء في ما يدفع ويقوّي وإمّا في الانتهاء في ما يحلّل. والقول المطلق في جميع الخنازير أنّ ما كان منها لا يجيب إلى التحلّل ينبغي أن يعفّن ويستأصل بالحديد ويحرز أن يصيب الحديد شيئا من العصب أو الشرايين أو العروق.

(٤٩) فصل: السلع ما كان منها رطوبة أرقّ وألطف كالعسلية فعلاجه بالأدوية القويّة التحليل وأمّا ما كان منها رطوبة أغلظ ينبغي أن يعفّن ثمّ يبطّ لأنّ التحليل لا

٦ يستفرغ] emend. eds. يستفراغ MS

correct [procedure] is to make an incision in the whole [skin covering the swelling] and to uncover it by lifting [the edge of the wound] with a probe; then one should remove the whole capsule [with the swelling]. If it is impossible to take it out completely, one should take out whatever one can and let the rest putrefy.

(50) If a soft swelling occurs incidentally following a disease, like that which occurs in the extremities [of the body] of those suffering from phthisis and dropsy, those extremities should be treated sometimes with rose oil, sometimes with olive oil and salt, and sometimes with vinegar and salt. If it is a disease originating from a small quantity of phlegm flowing to an organ, one should put a sponge that has been soaked in vinegar mixed with an equal amount of water on [the organ]; but if it originates from a large [quantity of phlegm], one should dissolve alum in water and vinegar, dip the sponge in [the mixture], and put it on [the swelling]. If the soft swelling originates from flatulence, it should be treated with sweet, strong medicines whose parts are fine and that dispel the winds (such as olive oil in which rue, lye from ashes, and vinegar have been boiled), as well as with the application of cupping glasses.

(51) [In the case of] bad tumors, take care to improve the temperament and humors of the body; and [in the case of] carbuncles, evacuate a large quantity of the patient's blood, as long as there is no hindrance. Then put a moderately dissolving and slightly repelling [remedy] on the surrounding area, and put on the [carbuncles] themselves remedies that have a strong drying effect and that have been mixed with wine that is either sweet, astringent, or something similar. Never apply a concocting medicine, so that [the carbuncles] do not putrefy. That would be very bad because carbuncles putrefy easily.

(52) When treating cancer, avoid remedies that produce black bile. Instead, institute a regimen for the liver [using] those ingredients that produce [healthy] blood and a [balanced] temperament, for this tumor originates from hot, melancholic blood. For [the same] reason—but also because [the liver] serves as the entry points of the veins—one should use those remedies that dry but do not burn, such as mineral remedies.

يكتفي به فيها. والصواب أن يبطّ جميعها وينبغي إذا بطّت أن يكشف عنها وترفع بالميل ويستقصى إخراج الكيس الذي هي فيه فإن عصي إخراج كلّه أن يخرج منه ما أمكن ثمّ يعفن الباقي.

(٥٠) فصل: الورم الرخو إن كان عرضا تابعا لمرض كالذي يعرض في أطراف ٥ المسلولين والمستسقين فعلاج تلك الأطراف مرّة بدهن ورد وأخرى بالزيت والملح وأخرى بالخلّ والملح. وإن كان مرضا حدث من انصباب بلغم إلى بعض الأعضاء فعلاجه إن كان يسير المقدار أن تحمل عليه إسفنجة قد شرّبت في الخلّ والماء ممزوجين باعتدال وإن كان كثيرا فيذاب الشبّ في الماء والخلّ وتغمس الإسفنجة فيه وتحمل عليه. وإن كان حدوث الورم الرخو عن ريح فعلاجه بالأدوية (fol. 117b) الحلوة ١٠ القوية اللطيفة الأجزاء المفشّة للرياح كالزيت المغلّى فيه السذاب وكماء الرماد والخلّ وكتعليق المحاجم.

(٥١) فصل: الأورام الرديئة المذهب ينبغي أن يعنى فيها بإصلاح مزاج البدن وأخلاطه فإنّ الجمر ينبغي أن يخرج لصاحبها من الدم مقدار كثير ما لم يمنع مانع ثمّ يوضع حواليها ما فيه تحلّل متوسّط ودفع يسير ويوضع عليها في أنفسها أدوية شديدة ١٥ التجفيف قد أديفت إمّا بشراب حلو أو عفص أو ما جرى مجراه. ولا ينبغي أن يستعمل فيها البتّة دواء منضج لئلا يعفن فيصير أردأ فإنّ الجمر سريع التعفّن.

(٥٢) فصل: السرطان يعالج باجتناب الأدوية المولّدة للكيموس السوداوي وبتدبير الكبد وبالأشياء المولّدة للدم والمزاج لأنّ هذا الورم يتولّد من دم حارّ سوداوي ومن أجل هذا ينبغي أن يستعمل فيه نفسه الأدوية التي تجفّف من غير

---

٥ تلك] emend. eds. ذلك MS || ٧ يسير] emend. eds. عين؟ MS || ١٣ وأخلاطه] emend. eds. وأخلطه MS || ١٥ عفص] emend. eds. عفس MS

If[63] one applies surgery, one should examine [the cancer] minutely and excise it completely so that no trace of it is left. One should let the blood flow and press and squeeze the surrounding area in order to expel the hot, thick, melancholic blood from it. One should not be quick to stop [the flowing of the blood]. Afterward, one should treat the wound appropriately.

(53) Scarification should be employed in order to either extract the bad matter that causes the organ to degenerate (but [only] after the body has been emptied) or facilitate the exit of the matter and its attraction [outward], as is done with cupping.

(54) Organs that have begun to putrefy and degenerate must be preserved, along with what surrounds them, by means of poultices and fomentations that preserve the innate heat and pneuma and gain strength for the organs, such as rose oil, saffron [*Crocus sativus* and var.], meal of bitter vetch [*Vicia ervilia*], wine, and the like.

(55) If part of the flesh of a nerve, ligament, or sinew degenerates, but its degeneration stops and it did not liquefy, one should then protect that which lies behind it, but it itself should [continue to] degenerate. Let it fall off [with the help of] offshoots of beets that are boiled and pounded with old butter and applied several times for a day and a night. If there is a small joint in that spot, one should rub it and [...]; if it is a bone, one should cut it with a saw, but be careful not to cut a healthy nerve. The degenerated nerve, on the other hand, should be putrefied with melted butter until it falls off.

(56) Cauterization should be applied (i) to stop [putrid] matter [from flowing], as when one cauterizes the head against the disease [called] elephantiasis; (ii) to prevent cankering and putrefaction, as when one cauterizes that which has become malignant; (iii) to dry the [putrid] matter and strengthen the organs, as when one cauterizes the spleen and the ischias; or (iv) to prevent blood from bursting forth, as when one cauterizes an artery.

تلذيع مثل المعدنية ولأنّه مداخل للعروق وينبغي إذا قطع بالحديد أن يستقصى قطعه والغوص عليه حتّى لا يترك له أثر ويترك الدم يسيل ويغمز على ما حواليه ويعصر لإخراج منه الدم الغليظ الحارّ السوداوي ولا يجل في حبسه ثمّ من بعد ذلك تداوى القرحة بما يوافقها.

٥ (٥٣) فصل: الشرط يستعمل إمّا في إخراج المادّة الرديئة المفسدة للعضو فيستعمل بعد استفراغ البدن ‹و›إمّا في تسهيل خروج المادّة وجذبها كما يعمل في الحجامة.

(٥٤) فصل: الأعضاء التي قد أخذت في العفن والفساد ينبغي أن تحفظ وما يجاورها بالأضمدة والنطولات التي تحفظ الحرارة الغريزية والروح وتكسب ١٠ الأعضاء قوة كدهن الورد والزعفران ودقيق الكرسنّة والشراب ونحو ذلك.

(٥٥) فصل: إذا فسد شيء من لحم العصب أو الرباط أو الوتر و وقف فساده و لم يذوب فينبغي أن يحفظ ما وراءه وأن يفسد هو نفسه ليسقط بأطراف السلق المسلوق المدقوق مع سمن عتيق ويلزم ذلك النهار والليل عدّة مرّات. فإن كان في ذلك الموضع مفصل صغير فافركه وارم(؟) به وإن كان عظما فانشره بمنشار وإيّاك ١٥ أن تقطع العصب الصحيح بل ينبغي أن يعفّن الفاسد بالسمن الذائب حتّى يسقط.

(٥٦) فصل: (fol. 118b) الكيّ يستعمل إمّا في قطع مادّة كما يكوى الرأس عن علّة الجذام وإمّا في منع التأكّل والعفونة كما قد يكوى ما قد تخبّث وإمّا في تجفّف المادّة وتقوية العضو كما يكوى الطحال وعرق النساء وإمّا في منع انبثاق الدم كما يكوى الشريان.

١٣ مرّات] emend. eds مرة MS || ١٨ انبثاق] emend. eds. انبثق MS

(57) All ulcers should be made to possess angles [after surgical removal]. And if one has to hollow them out, one should do so in a form appropriate to this concern, so that nature will have appendages from which she can start to grow flesh. The more angular the ulcer is, the easier it is to heal. Roundness hinders fast formation of flesh for three reasons: (i) the fibers are cut off in their length and breadth [crosswise], (ii) two [double?] bandages cannot unite the two margins, [and] (iii) nature does not find a spot from where she can start to grow flesh.

(58) The [successful] treatment of an ulcer is achieved through [the observation of] seven things:

i. The kind of ulcer: From this, one can learn what to do; for if it is simple, one needs only to close it up. If flesh has been destroyed, there is a need for [new] flesh to be grown in order to replace the other flesh; then the ulcer needs to be sealed and form a scar. If matter flows toward it, that flow needs to be stopped, and then it needs to form a scar. If ugly flesh grows in it, that flesh needs to be extirpated through surgery and drugs that eat away at it; then [the ulcer] needs to be sealed and to form a scar. If it is accompanied by a bad temperament, that [temperament] needs to be eliminated; then [the ulcer] needs to be sealed and to form a scar. If the ulcer is composed of many things, the treatment should be put together according to all these things and each element applied according to its importance.

ii. [Its] cause: [If it is caused] by the bite of the poisonous animal, the poison should be countered first of all; it should be drawn [outward] and [its effects] moderated. Often one needs to widen the ulcer, as in the case of the bite of a mad dog.

iii. The part of the body where the ulcer is: Each organ requires a specific treatment.

iv. The temperament, strength, external condition [of the body], and habits [of the patient]. A dry body needs medicines that are more dry, and a moist body needs drugs that are more moist, since the aim [of the treatment] is the restoration of

(٥٧) فصل: القروح كلّها ينبغي أن تجعل ذوات زوايا وإن اضطررت إلى تقويرها فاجعل ذلك على شكل موافق للأمر لتكون للطبيعة زوائد تبتدئ منها بإنبات اللحم وكلّما كانت القرحة أكبر زوايا كانت أسهل برءا وذلك أنّ الاستدارة تعوق عن سرعة الالتحام من قبل ثلاثة أشياء: ١) أنّ الليف قد انقطع طولا وعرضا ٢) أنّ الرفادتين لا ٥ يمكن أن تجمع شفاها ٣) أنّ الطبيعة لا تجد موضعا تبتدئ منه بانتبات اللحم.

(٥٨) فصل: علاج القروح يكتسب من ز أشياء: ١) نوع القرحة فإنّ منه يعلم ما ينبغي أن يفعل وذلك أنّها إذا كانت بسيطة فالحاجة إلى إلحامها فقط وإن كان قد ذهب منها لحم فالحاجة إلى أن ينبت فيها اللحم عوضا من الآخر ثمّ إلى إختامها وإدمالها وإن كان ينصبّ إليها مادّة فالحاجة إلى قطع الانصباب ثمّ إلى الاندمال ١٠ وإن كان قد ينبت فيها لحم سمج فالحاجة إلى استئصال ذلك اللحم بالحديد والأدوية الأكلة ثمّ إلى الختم والإدمال وإن كان معها سوء المزاج فالحاجة إلى إزالته ثمّ إلى الختم والإدمال وإن كان تركيبها من أشياء كثيرة فالحاجة أن يؤلّف العلاج بحسب جميعها وتقدّم الأهمّ في الأهمّ. ٢) من السبب: فإن نهشة الحيوان المسموم ينبغي أن يقاوم أوّلا ذلك السمّ ويجتذب ويعدّل وكثيرا ما يحتاج فيها إلى توسيع القرحة ١٥ في نحو عضّة الكلب الكلب. ٣) من العضو الذي (fol. 118b) فيه القرحة فإنّ كلّ واحد من الأعضاء يحتاج إلى علاج خاصّ. ٤) من المزاج والقوة والسحنة والعادة وذلك أنّ البدن اليابس ينبغي أن تكون أدويته أشدّ يبسا والبدن الرطب يحتاج

١ زوايا] .emend. eds زويا MS || ٢ موافق للأمر] .emend. eds و.ق الأمر؟ MS || ٣ زوايا] .emend. eds زويا MS || ٤ الرفادتين] .emend. eds الرفائدان MS || ١٧ يحتاج] قروحه add. MS[1]

continuity. And a strong body endures the kind of heavy treatment that a weak [body] does not endure.

v. The things that, through their conformity or lack thereof, are an indication for [the kind of treatment] needed. In the case of summer and a hot country, the temperament requires medicines that are less heat-producing.

vi. The powers of the medicines and foods, so that one can select from them that which is best suited for reaching the goal one aims at.

vii. The comparison of some of these [points] with others so that, from all of them, one can [decide upon] the [right] kind of treatment.

(59) Medicines that make the flesh grow in [place of an] ulcer are those that preserve its temperament, that clean and dry as much foul matter as originates in that ulcer. If their cleansing effect is too strong, they annihilate the matter from which flesh originates; and if their drying effect is too strong, they make the ulcer form a scar and prevent the growth of flesh. If they are too hot or not hot enough, they produce a bad temperament in that part of the body, which [in turn] sometimes causes something else. The medicines that cause an ulcer to close are those that dry the temperament of the body and its parts to the second degree, so that the [foul] matter dries up and becomes hard according to their dryness. Medicines that make an ulcer form a scar are those that dry the temperament of the body to the third degree, so that their intense dryness makes the nature of the flesh hard.

(60) Ulcers may be hard to heal because of a bad temperament of the flesh in which they are [found], or because of the bad quantity or quality of the blood that reaches them, or because of the kind of drugs [administered], or because of their bad arrangement,[64] or because of a bad treatment. The indications for a bad temperament of the flesh that is hot are the heat of that spot, its red or black appearance, the opacity of the ulcer, the large quantity of thin pus within, [and the fact that] the patient feels a burning heat in it but finds relief in cool things. The indications

إلى أدوية أرطب من قبل أنّ الغرض في ذلك هو ردّ الاتصال وأيضا فإنّ البدن القوي يحتمل من العلاج الصعب ما لا يحتمله الضعيف. ٥) من الأشياء التي باتّفاقها واختلافها يستدلّ على ما يحتاج إليه فإنّ الصيف والبلد الحارّ والمزاج يحتاج فيهما إلى أدوية أقلّ إسخانا. ٦) من قوى الأدوية والأغذية لتختار منها الأوفق ٥ بالأوفق في البلوغ إلى الغرض المقصود. ٧) من مقايسة هذا بعضها ببعض ليلتمّ من جميعها نوع العلاج.

(٥٩) فصل: الأدوية التي تنبت اللحم في القروح هي التي تحفظ عليه مزاجه وتجلو وتجفّف بمقدار ما يتولّد في القرحة من الوضر والوسخ فإنّه إن زاد جلاها أفنت المادّة التي يتولّد منها اللحم وإن زاد تجفّفها أدملت القرحة ومنعت من إنبات اللحم وإن زاد ١٠ إسخانها أو نقص أحدثت في العضو سوء مزاج ربّما كان سببا لشيء آخر. والأدوية التي تلحم القروح هي التي تجفّف في ب من مزاج البدن ومن مزاج العضو لتنشّف المادّة وتصلب بقدر تجفّفها والأدوية الداملة هي التي تجفّف في ج من مزاج البدن لتبلغ شدّة تجفّفها إلى أن يخرج اللحم من طبيعته إلى الطبيعة الصلبة.

(٦٠) فصل: القروح يعسر برؤها إمّا من قبل سوء مزاج اللحم الذي هي فيه وإمّا ١٥ من قبل رداءة الدم بجيئها في كميته أو كيفيته وإمّا من قبل نوع الأدوية وإمّا من قبل سوء ترتيبها وإمّا من قبل رداءة العلاج. وسوء مزاج اللحم إن كان حارّا فعلامته حرارة الموضع وحمرة اللون أو سواده وغلظ شفاف القرحة وكثر الصديد الرقيق فيها وأن يحسّ العليل اللهب فيها ويستريح إلى الأشياء الباردة وإن (fol. 119a)

٤ لتختار] .emend. eds لتختر MS || ٨ أفنت] .emend. eds أفنات MS

for a cold [temperament] are coldness of the spot; a white, green, livid, or leaden [color]; its lack of pus; [and the fact that] the patient feels coldness in it and finds relief in hot things. The indications for a moist [temperament] are severe flabbiness and much foul matter. The indications for a dry [temperament] are intense dryness and bad blood.

(61) The difference between a swelling occurring from the dissolution of soft flesh by very hot drugs and [the swelling] occurring from the flow of matter [to a certain organ] is that the dissolution of the flesh is accompanied by an increase in heat, redness, and hardness of the lips of the ulcer, and heat of the surrounding area. [Another difference is] that the foul matter occurring from the dissolution of the flesh is thin and red or tending to lividity, while that occurring from the flow of matter is thick and viscous.

(62) Some remedies for ulcers cleanse and dry at the same time, such as frankincense [*Boswellia carterii* and var.], sarcocol [gum resin of *Astragalus sarcocolla*], aloe [*Aloe vera*], lycium juice, opopanax [*Opopanax chironium* Koch], root of Florentine iris [*Iris florentina*], meal of bitter vetch, birthwort [*Aristolochia clematis*], myrrh [*Commiphora myrrha*], and resin. Others dry and cool, such as gallnuts [of *Quercus* sp. and var.], pomegranate peels and the blossom of the wild pomegranate tree, carob [*Ceratonia siliqua*], and Maltese mushroom [*Cynomorion coccineum*]. Yet others dry and make the flesh grow, such as Armenian clay and dragon's blood [red resin of *Dracaena draco*]. Some are minerals, such as cadmia, tutty, ceruse, litharge, vitriol, and alum. Others are liquid, such as vinegar and rose oil. For every situation, one should select from among these the ones that are most appropriate.

(63) One should select remedies according to the difference in the organs. For ulcers in the digestive organs, it is appropriate [to apply] heat as well as astringent herbs and what is prepared from them, such as rhubarb-currant [*Rheum ribes*], myrtle [*Myrtus communis*], sumac, blossom of the wild pomegranate tree, goatsbeard [*Tragopogon pratensis* or *Tragopogon porrifolius*], pomegranate peels, acacia [the sap of the acacia tree, *Acacia arabica* Willd. or *Acacia nilotica* Del.—that is, gum arabic], quince [*Cydonia oblonga*], apple, wine, and gallnut. The remedies for ulcers in the respiratory organs should be mixed with mineral [substances] that alleviate cough, such as gum tragacanth, gum [arabic],

كان باردا فعلامته برد الموضع وبياضه وخضره وكمودته و رصاصيته وقلّة الصديد وأن يحسّ العليل البرد فيها ويستريح إلى الأشياء الحارّة وإن كان رطبا فعلامته شدّة الترهّل وكثرة الوضر وإن كان يابسا فعلامته شدّة القحل و رداءة الدم.

(٦١) فصل: الفرق بين الرهل الحادث من ذوبان اللحم الرخص الذي يكون من قبل ٥ الأدوية القوية الحرّ وبين الذي يكون من انصباب مادّة هو أنّ ذوبان اللحم يحصل معه زيادة في سخونة القرحة واحمرار وصلابة شفاهها وحرارة ما حولها وأيضا فإنّ الوضر الكائن عن ذوبان اللحم رقيق أحمر أو مائل إلى الكمودة والوضر الكائن عن انصباب المادّة غليظ لزج.

(٦٢) فصل: أدوية القروح منها ما تجلو وتجفّف معا كالكندر والأنزروت ١٠ والصبر والحضض والجواشير وأصل السوسن الأسمانجوني ودقيق الكرسنّة والزراوند والمرّ والراتينج ومنها ما تجفّف وتبرّد كالعفص وقشور الرمّان والجلّنار والخروب والطراثيث ومنها ما تجفّف وتلحم كالطين الأرميني ودم الأخوين ومنها معدنية كالإقليميا والتوتيا والإسفيداج والمرتك والزاج والشبّ ومنها ماعية كالخلّ ودهن الورد. وينبغي أن يختار منها في كلّ موضع ما هي أوفق.

١٥ (٦٣) فصل: ينبغي أن تختار الأدوية بحسب اختلاف الأعضاء فإنّ قروح آلات الغذاء يوافقها الإسخان والحشائش القابضة وما يتّخذ منها كالريباس والآس والسمّاق والجلّنار ولحية التيس وقشور الرمّان والأقاقيا والسفرجل والتفّاح والشراب والعفص. وقروح (fol. 119b) آلات التنفّس يحتاج أن يخلط في أدويتها الأدوية

٤ الحادث] .emend. eds الحديث MS || ١٢ الأخوين] .emend. eds الأخاوين MS || ١٣ والزاج] .emend. eds والزج MS || ١٧ والجلّنار] .emend. eds والجلّبنار MS

and starch. For tumors in the brain, one should use [the different kinds] of oil and their compositions. It is clear that Armenian earth and similar [ingredients] are appropriate for all the other ulcers that are accompanied by inflammation.

(64) Ulcers in the intestines and upper part of the stomach should be treated by taking appropriate medicines and foods. Those in the large intestines should be treated by applying clysters and suppositories. Those in the respiratory organs should be treated by [remedies] that are kept in the mouth or ingested by drinking or inhalation. Those in the mouth should be treated by gargling. Those in the esophagus and stomach—that is, the cardia [of the stomach]—require agglutinants that adhere to the spot. Moreover, the stomach needs [remedies] that stay there for a long time; and the remote organs, such as the lungs and the kidneys, need [remedies] that quickly get through to them and reach them. One should determine the requirements of each organ. One should also apply all the [appropriate] strong poultices.

(65) [Further] examples of what the organs need: Ulcers in the nose need remedies that are more drying than [those needed for] tumors in the eyes. Ulcers in the ears need remedies that are more drying than [those needed for] ulcers in the thighs, because the ears are [naturally] drier. They also need more honey, since more foul matter originates in them. Ulcers in the mouth need juice of unripe, sour grapes; sumac; or a rob of mulberries or walnuts. If they are accompanied by putrefaction, they need caustic remedies with a strong drying effect. Ulcers in the penis, vagina, and anus always need extremely drying [remedies], such

المعدنية المسكّنة للسعال كالكثيراء والصمغ والنشاء وقروح الدماغ تحتاج إلى أدهان
وما يتّخذ منها. وظاهر أنّ الطين الأرماني وما جرى مجراه يوافق بسائر القروح التي
معها تلذّع.

(٦٤) فصل: ما كان من القروح في الأمعاء والمعدة العليا مداواتها بالأدوية
٥ والأغذية الموافقة بالأكل والشرب وما كان منها في الأمعاء الغلاظ فمداواتها
بالحقن والتحميل وما كان منها في آلات التنفّس فمداواتها بما يمسك في الفم وبما
يشرب وبما يصل بالتنفّس وما كان منها في الفم فمداواتها بما يتغرغر به وما كان منها
في المريء والمعدة أعني فمها فهي تحتاج إلى ما يغرى أو يلصق بالموضع وأيضا فإنّ
المعدة تحتاج إلى ما يطول لبثه فيها والأعضاء البعيدة كالرئة والكلى تحتاج إلى ما
١٠ يسرع نفوذه ووصوله إليها. وينبغي أن يقدّر لكلّ عضو ما يحتاج إليه وأن يستعمل
الأضمدة القوية فيها كلّها.

(٦٥) فصل: المثال فيما يحتاج إليه الأعضاء: قروح الأنف تحتاج إلى أدوية
أكثر تجفّفا من قروح العين وقروح الأذان تحتاج إلى أدوية أكثر تجفّفا من قروح
الفخذين لفضل يبس الأذن على هذا وتحتاج أيضا إلى فضل عسل لفضل ما يتولّد فيها
١٥ من الوضر وقروح الفم تحتاج إلى عصارة الحصرم والسمّاق وربّ التوت وربّ
الجوز فإن كان معه عفونة فهي تحتاج إلى أدوية قوية التجفّف والكاوية وقروح
الإحليل والفرج والمقعدة تحتاج أبدا إلى ما هو أشدّ تجفّفا كالقرطاس المحروق

١ كالكثيراء والصمغ والنشاء وقروح الدماغ تحتاج إلى أدهان وما يتّخذ منها MS[1] || ٧ فمداواتها] emend. eds. فمداوتها MS || ١٠ وأن يستعمل] emend. eds. واستعمل MS || ١٤ الفخذين] emend. eds. الفخذان MS

as burned papyrus and burned seashells and the like. If they are recent, [one should treat them with] aloe and lycium juice and the like.

(66) When foul matter accumulates in ulcers, they often need to be washed. If they are in the external parts and have a cold temperament, [they should be washed] with hot wine; if they are hot, with vinegar mixed in sweet water and with water in which poppy [*Papaver somniferum* and var.] has been cooked; if they are moist, with water in which pomegranate peels and myrtle have been cooked; and if they are dry, with sweet water. If there is an ulcer in the internal parts, [it should be washed] with sugar water and hydromel and milk.

(67) Ulcers in the lungs are most difficult and hard to cure because medicines reach them only after their potency has been weakened and because the lungs are organs that move constantly. Movement prevents the parts of the ulcer from joining together, and it does not allow the medicine to adhere. For this reason, it is not easy to cure them.

(68) If a dissolution of continuity[65] occurs in the lungs, either because of a torn vein or because of something else, one should immediately begin with treatment by telling the patient to rest, to speak little, and to refrain from coughing as much as possible; by attracting the [purulent] matter from the lungs through bleeding the basilic vein and bleeding [the patient] many times; by rubbing his hands and feet; and by extracting blood resulting from astringent, glutinous food on the [same] day, if the strength of the patient allows it. One should apply a poultice with strong, restraining medicines to the chest, such as rose and quince oil in the summer and nard oil in the winter.

والودع المحروق وما أشبه ذلك وأيضا إن كانت قريبة العهد فالصبر والحضض وما أشبه ذلك.

(٦٦) فصل: القروح التي تجتمع فيها وضر كثيرا ما تحتاج أن تغسل فإن كانت في الأعضاء الظاهرة وكانت باردة المزاج فبالشراب السخن وإن كانت حارّة فبالخلّ

٥ (fol. 120a) الممزوج بالماء العذب وبماء طبخ فيه خشخاش وإن كانت رطبة فبماء طبخ فيه قشور رمّان وآس وإن كانت قحلة فبالماء العذب وإن كانت القرحة في الأعضاء الباطنة فبماء السكّر وماء العسل واللبن.

(٦٧) فصل: أصعب القروح وأعسرها برءا قروح الرئة من قبل أنّ الأدوية لا تصل إليها حتّى تضعف قواها ولأنّ الرئة عضو دائم الحركة والحركة تمنع اتّصال

١٠ أجزاء القرحة ولا تترك الدواء يلزمها فلهذا السبب ليس بسهل برؤها.

(٦٨) فصل: إذا حدث تفرّق الاتّصال في الرئة إمّا من انخراق عرق أو غيره فينبغي أن تأخذ في علاجه من ساعته بأن تأمر العليل السكون وقلّة الكلام والتشاغل عن السعال ما أمكن وأن تجذب المادّة من الرئة بفصد الباسليق وإخراج الدم في دفعات كثيرة وبأن تدلك اليدين والرجلين وإن كانت قوة العليل فيما تحتمل بإخراج

١٥ الدم أيضا من اليوم من الغذاء قابضا لزجا وضمّد الصدر بأدوية قوية رديعة إمّا في الصيف فكدهن الورد والسفرجل وإمّا في الشتاء فكدهن الناردين.

١ والودع] emend. eds. والوضع MS || ٣ فيها] emend. eds. فيه MS || ٤ حارّة] emend. eds. حراة MS || ٥ خشخاش] emend. eds. خشخش MS || ٩ إليها] emend. eds. التي MS | الحركة] emend. eds. الحراكة MS | والحركة] emend. eds. والحراكة MS || ١٥ لزجا] emend. eds. لزجه MS

(69) Internal ulcers need drying medicines, such as the theriac,[66] yellow amber [resin of *Populus nigra*] pills, coral, Armenian earth,[67] sealed earth [*terra sigillata*], Indian nard [*Nardostachys jatamansi*], and saffron. For ulcers in the lungs, one should select the strongest among them, adding some opium in order to alleviate the cough, stop the catarrh, bring about sleep, [and increase] the drying effect. Foods should be those that are quickly digested and that are astringent and viscous, such as [the dish called] trotters[?],[68] barley broth, small fishes cooked with quince and apples, and the like. In the same manner, one should select the appropriate foods and medicines for every ulcer.

(70) If an ulcer occurs in the lungs because of a hot defluxion in the head, the defluxion should be stopped. The head should be shaved; and one should put vesicatory remedies on it, such as the mustard remedy, so that the vapor that is congested in the head is dispersed and dissolved. Be careful that no oil [or no] hot or severely cold ingredient reaches the head. Give [the patient] remedies that dry the ulcer and stop the catarrh, such as the theriac. If he needs to be taken to the bathhouse in order to ripen the catarrh, this should be done. Give him a refining treatment so that the catarrh does not increase.

(71) Fundamental for returning the lungs to their normal condition are [the following]: alleviating the cough; stopping the catarrhs; removing the bad humors in the body; taking beneficial medicines and keeping them in the mouth [for a while]; lying on one's back; relaxing the muscles of the throat so that the medicines go down to the lungs bit by bit; exerting oneself to purge [the body] with emollient medicines that expel the superfluities [from it]; balancing the temperament; and giving [the patient] soft foods prepared from starch, bean flour, and barley flour.

(٦٩) فصل: القروح الباطنة تحتاج إلى الأدوية المجفّفة كالترياق وأقراص الكهرباء والبسد والطين الأرماني والمختوم والسنبل والزعفران وينبغي أن يختار منها لقروح الرئة ما كان أقوى وقد حصّل فيه شيء من أفيون لإسكان السعال ولقطع النزل ولجلب النوم مع تجفّفه. وأمّا الأغذية فينبغي أن تكون (fol. 120b)
٥ سريعة الانهضام وفيها قبض ولزوجة كالأكارع(؟) وحسو الشعير والسمك الصغير إذا طبخ فيه السفرجل والتفّاح وما جرى مجراها وعلى هذا القياس ينبغي أن يختار لكلّ قرحة ما يوافقها من الأغذية والأدوية.

(٧٠) فصل: إذا حدثت قرحة في الرئة عن نزلة حارّة في الرأس فينبغي أن تمنع النزلة وأن يحلق الرأس وأن توضع عليه الأدوية المنفّطة كدواء الخردل ليفشّ ويحلّل
١٠ عن الرأس ما احتقن من البخار واحذر أن يصيب الرأس شيء من الدهن أو من الحارّ أو من البرد الشديد. وأعط من الأدوية ما يجفّف القرحة ويمنع من النزل كالترياق وإن احتاج إلى إدخال الحمّام لإنضاج النزلة فافعل ولطّف لئلا تزيد النزلة.

(٧١) فصل: ملاك الأمر في ردّ الرئة إلى الاعتدال يكون بسكون السعال ومنع النزلات وإزالة الأخلاط الرديئة التي في البدن وشرب الأدوية النافعة وأخذها
١٥ في الفم والاستلقاء على القفاء وإرخاء عضل الحلق لينزل من الأدوية شيء بعد شيء إلى الرئة وتجاهد الإسهال بالأدوية الليّنة التي تخرج الفضول وتعدّل المزاج وأعطهم الأغذية الليّنة التي تعمل من النشاء ودقيق الباقلى والشعير.

---

٢ الكهرباء] emend. eds. الكرابا MS || ٣ لإسكان] emend. eds. لإسكن MS || ٤ ولقطع] emend. eds. ويقطع MS | ولجلب] emend. eds. ويجلب MS || ١١ وأعط] emend. eds واعطي MS || ١٢ احتاج] emend. eds. احتاجت MS | لإنضاج] emend. eds. لإنضج MS || ١٣ ردّ] emend. eds. رادّ MS || ١٦ تخرج] emend. eds. تخراج MS || ١٧ وإعطهم] emend. eds. واعطيهم MS

(72) [When blood is discharged from] the lungs, one should administer to the patient vinegar mixed with a lot of water, hour after hour, in order to stop [the discharge of] coagulated blood. If there is an ulcer in the lungs or chest and it is accompanied by a rattling sound in the chest and by difficulty breathing, one should administer to the patient cleansing agents that ease the discharge of the [feculent] matter, such as honey water, barley gruel, julep, or soup. If the [feculent] matter is thin, [the patient] should be given a thickener, such as starch, so that the [subsequent] discharge by means of honey becomes easier. If, in the case of ulcers in the lungs, one gives the patient wine to drink, one should mix it with water in which quinces or myrtle seed or other astringent ingredients have been cooked, in order to strengthen the lungs and dry the ulcer.

(73) Diverting [surplus] matter to the opposite side is accomplished by bleeding from the opposite side, extracting the blood therefrom many times, and applying cupping glasses. In the same manner, if we see a nosebleed that increases, becoming more severe, which cannot be stopped with remedies, we bleed from the veins of the arm that is on the same side as the nostril from which the blood flows, and we apply cupping glasses to the hypochondria. If the blood flows from the right nostril, [we apply cupping glasses] to the liver; and if it flows from the left nostril, [we apply them] to the spleen so that the blood is attracted downward. We also tie the hands and feet tightly [with bandages] once they have been rubbed.[69]

(74) A hemorrhage is stopped by astringent remedies that are cohesive and agglutinant. The best are those that make the flesh grow; the faster it grows, the better it is.

(75) If you want to stop a hemorrhage, you should first of all put a medicine on it that stops the bleeding; then apply a bandage and observe: if it still bleeds a little bit, you should not untie the bandage; but if it bleeds a lot, you should untie the first bandage and repeat this procedure after you let the patient rest for a little while. You should leave on the bandage until the third day, then untie it gently, and if you find the medicine [still] adhering to the wound, you should add [some more medicine] to it and put a [new] bandage on it. And if you find that [the wound] has healed, you should remove it [the bandage] gently and do the same thing again.

(٧٢) فصل (fol. 121a): ‹. . .› الرئة ينبغي أن يسقى العليل خلّ ممزوج بماء كثير ساعة بعد أخرى لينقطع ذلك الدم المنعقد وإذا كان في الرئة أو في الصدر قرحة وحدث معها جرجرة في الصدر وضيق نفس فأعطه ما ينظّفه ويسهّل خروج تلك المادّة كماء العسل وماء الشعير أو الجلاب أو الحساء وإن كانت المادّة رقيقة فأعط
٥ ما يغلّظها ليسهل خروجها بالعسل كالنشاء وإذا سقيت شرابا في قروح الرئة فاجعله بماء قد طبخ فيه سفرجل أو حبّ الآس أو غيره من الأشياء العفصة لتقوّى الرئة وتجفّف القرحة.

(٧٣) فصل: إمالة المادّة إلى الجانب المخالف يكون بفصد الجهة المخالفة وإخراج الدم منها في دفعات كثيرة وتعليق المحاجم عليه. ‹و›مثل ذلك إذا رأينا الرعاف قد يزيد
١٠ ويفرط ولم يسكن بالأدوية فصدنا عروق اليد في جانب المنخر المرعوف ووضعنا المحاجم على مراقّ البطن. وأمّا إن كان الرعاف من المنخر الأيمن فعلى الكبد وإن كان من المنخر الأيسر فعلى الطحال ليجذب بذلك الدم إلى أسفل وشدّ اليدين والرجلين شدّا عنيفا وتدلّكها.

(٧٤) فصل: الأدوية القاطعة انبثاق الدم هي القابضة التي تلصق وتغري
١٥ وأجودها ما كان معه إنبات اللحم وكلّما كان إنباته أسرع كان أجود.

(٧٥) فصل: إذا أردت أن تشدّ انبثاق الدم فاجعل عليه أوّلا دواء قاطعا للنزف ثمّ اربطه وانظر فإن انبعث منه دم يسير فلا تحلّه وإن انبعث منه دم كثير فحلّ الرباط الأوّل وأعد العمل بعد أن تريح العليل هنيهة(؟) ودع الرباط إلى اليوم الثالث ثمّ حلّه برفق فإن وجدت الدواء لاصقا بالجرح فزد عليه ثمّ اربطه فإن وجدته قد تبرّأ
٢٠ فاقلعه برفق وأعد العمل الأوّل.

---

٤ وأعط] emend. eds واعطي MS || ٩ عليه] ودلكه add. MS || ١٩ فزد] emend. eds. فزاد MS

(76) If the blood is not contained with the bandage, one should scrape off the skin and catch the vein on a hook, pull it upward, and bind both edges with a silken thread while the vein is in the middle. When there is a hemorrhage, be careful with cauterization. Although cauterization stops the hemorrhage, it is not safe, because it takes away from the flesh of the spot. The bleeding is indeed contained as long as there is a scab on the spot [of the wound], but if the scab falls off, a [subsequent] bleeding is more difficult to treat. Apply cauterization only when it is necessary. Prefer caustic medicines to cauterization. Cauterization is good when bleeding is associated with putrefaction and when it concerns [wounds that] putrefy quickly, such as [those in] the anus and testicles. Caustic medicines that are beneficial for [these wounds] are arsenic, verdigris, lime, sulphur, green vitriol, and the like.

(77) Wounds are either harmless or dangerous. For instance, wounds that affect the chest cavity or the digestive organs are dangerous. Equally [dangerous are wounds] that occur to nervous parts such as the diaphragm, the joints, and the nerves because they induce spasms and delirium, and [wounds] that occur to the blood vessels because they cause bleedings. A wound that occurs in very fleshy areas is more dangerous to [those areas]. And wounds that occur to the large intestines are easier to heal than those that occur to the small intestines, especially the jejunum. If it is torn, it is almost completely impossible to cure because of its thin substance, which is similar to the nature of the nerves.[70]

(78) If much blood flows from a wound right away, it is less dangerous than a wound from which no blood flows. The reason is that this [latter] wound is close to being a tumor, from which no blood flows, while [the wound] from which blood flows is not close to being a tumor. For this reason a wound from which no blood flows requires venesection.

(٧٦) فصل (fol. 121b): إذا لم يمتسك الدم بالرباط فاقشط الجلد وعلّق(؟) العرق بصنارة واجذبه إلى فوق واربط بخيط إبريسم الجانبين والعرق في الوسط واحذر في نزف الدم الكيّ بالنار وذلك أنّ الكيّ وإن سكّن انبثاق الدم غير مأمون لأنّه قصّ من لحم الموضع وإنّما يمسك الدم بإمساك قشرة على الموضع فإذا عرض ٥ أن تسقط القشرة كان نزف الدم أصعب وأعسر علاجا ولا تفعل الكيّ إلا عند الضرورة واختر الأدوية الكاوية على الكيّ والكيّ خير فيما كان من النزف مع عفونة والسريعة العفن كالدبر والمذاكير والأدوية الكاوية التي تصلح لذلك الزرنيخ والزنجار والنورة والكبريت والقلقطار وما أشبه ذلك.

(٧٧) فصل: الجراحات منها سليم ومنها مخوّف. مثال ذلك أنّ الجراح التي ١٠ تبلغ تجويف الصدر أوآلات الغذاء خطرة وكذلك التي تقع في الأعضاء العصبية كالحجاب والمفاصل والأعصاب لأنها تجلب التشنّج واختلاط العقل وكذلك التي تقع في العروق لأنّها تحدث نزف الدم والجراح التي تقع في المواضع الكثيرة اللحم أخطر فيها وأيضا الجراح التي تقع في الأمعاء الغلاظ أسهل برء من التي تقع في الأمعاء الدقاق ولا سيّما الصائم فإنّه إذا تخرّق لا يكاد يبرأ أصلا لرقّة جرمه وقربه ١٥ من طبيعة العصب.

(٧٨) فصل: الجراح الذي يسيل منه من أوّل الأمر دم كثير أسلم من الذي لا يسيل منه دم وذلك أنّ هذا على قرب من الورم الذي ليس يجري منها دم والذي يجري منه دم بعيد من الورم ومن أجل هذا تحتاج الجراح التي لا ينبعث منها دم الفصد.

---

٣ الكيّ] .emend. eds القيّ MS | انبثاق] .emend. eds انبثق MS | الدم] من add. MS || ٨ والقلقطار] .emend. eds والقلقطر MS || ١٣ من] ditt. MS

(79) If there is a wound to the lower [part of the] belly and the omentum protrudes and some of it turns green or black, then do not hope [to heal] that which has turned green. Instead, after you have come to the aid of the healthy [part], cut off [the diseased part] by ligating it with a fine silken thread, so that the cut does not result in a hemorrhage from the blood vessels in the omentum. Let the ends of the thread come out from the wound so that it will be possible to extract it if the wound closes up.[71] If the gut protrudes from a wound and becomes inflated and cold because of the air, place on it a hot compress with astringent, hot wine so that it is warmed and its strength is preserved. If that is not possible, take hot [pieces of cloth?] and put [them] on [the gut] in the bathhouse, or put a sponge or piece of wool [that has been soaked] in hot olive oil or hot water on it so that it returns to its [former] condition.[72]

(80) If the omentum or intestines protrude, raise the patient by his hands and feet, in hot air and in a manner such that the abdomen is drawn upward and becomes clearly visible, while the organ does not become cold. Then the patient should be softly shaken and gently put to sleep in this position on a flat bed raised at its extremities. Once the patient has been put to sleep and tied up in the middle [?], one should make efforts to return [the omentum or intestine] inside [the body]. Then one should draw the edges of the wound together and cut it[73] carefully. If the patient needs to defecate, give him a clyster with [ingredients] that expel the feces, and alleviate the pain with astringent black wine or the like.

(81) If one needs [to] suture [the wound], one should bring both edges of the wound together and stitch them, then fasten the thread with a double knot and cut it off. Then skip over a small [part?] of the wound, join the two edges, and stitch them in the same way. Continue to operate in this manner until you reach the end of the wound. The threads should consist of [material] that does not decay quickly.

(٧٩) فصل: إذا وقع الجرح في البطن الأسفل ويبرز الثرب فاخضرّ منه شيء أو اسودّ فلا تطمع في ما اخضرّ منه ولاكن اقطعه بعد أن يستوثق من الصحيح بالشدّ (fol. 122a) بخيط دقيق من إبريسم لئلا يحدث من القطع نزف الدم من العروق التي في الثرب واجعل طرفي الخيط خارج من الجرح ليمكن أن يخرج إذا التحم وإذا برز
٥ المعاء من جرح وانتفخ وبرد من الهواء ينبغي أن يكمّد بشراب قابض مسخّن ليسخّنه ويحفظ عليه قوته فإن لم يتهيأ ذلك فبربات(؟) حارّة تلزق عليه في الحمّام أو يكمّد بدهن مسخّن أو بالماء الحارّ بإسفنجة أو صوفة ليعود إلى حاله.

(٨٠) فصل: إذا برز الثرب أو أمعاء من خارج ينبغي أن يعلّق العليل بيديه ورجليه في هواء حارّ لينجذب ويتوضّح بطنه ولا يبرد العضو ثمّ يهزّ هزّا رفيقا وينوّم
١٠ العليل برفق على هذا الهيئة فوق فراش وطيء مرتفع الأعالي وينوّم العليل متعقّد(؟) الوسط ويحتال في ردّه إلى داخل ثمّ ضمّ شفتا الجرح وقطّه(؟) بإحكام وإن احتاج العليل إلى إخراج الزبل فاحقنه بما يخرج الثفل ويسكّن الألم بمثل الشراب الأسود القابض.

(٨١) فصل: إذا احتجنا إلى الخياطة ينبغي أن يجمع طرفي الجرح وتخطّهما وتعقد
١٥ الخيط عقدتين ثمّ تقطعه وتترك مقارن(؟) يسير من الجرح وتجمع الطرفين وتخطّهما كالأوّل ولا تزال تفعل ذلك إلى أن يبلغ إلى أمر الجرح وينبغي أن تكون الخيوط ممّا لا يسرع إليه العفن.

---

٤ طرفي] emend. eds. طريف MS || ٨ بيديه] emend. eds. بيده MS || ٩ ويتوضّح] emend. eds. ويتقطّع؟ MS || ١١ شفتا] emend. eds. شفا MS

(82) Puncturing a nerve causes severe pain that attracts matter to the spot. And if this results in a tumor in that spot, one should leave the tear open and widen it if it is narrow. One should do what one can to alleviate the pain through rubbing and plastering with oils and hot, fine ingredients that have a heating effect but are not astringent. One should beware of fomentation and bathing in hot water because it has an excessive moistening effect. Since the substance of the nerves is moist and solid because [nerves are by nature] cold, [the nerve] would putrefy quickly if it were moistened and heated.

(83) Remedies appropriate for tumors in the nerves are those that are lukewarm, consist of fine parts, and are neither astringent nor pungent, such as sweet olive oil, mastic from the turpentine tree, asafetida [gum resin of *Ferula asafoetida*], opopanax, and spurge [*Euphorbia resinera* and var.]. Select the weakest of these for [patients] with moist bodies. When the nerve is exposed, the power of the medicine reaches it [easily]. [And select from] the strongest medicines for [patients] with a dry body. When the nerve lies deep within, the power of the medicine cannot reach it. Tumors of the nerves need medicines that are moderately hot and intensely dry and that have[?] a fine substance. If there is a puncture in the arm, rub the neck and armpits with hot oil; if it is in the leg, rub the groins [with this medicine]. Medicines for all tumors in the nerves should be calefacient, actually, so that their [the nerves'] coldness does not induce spasms. One should pour oil on an abscess so as not to increase its foulness. If a contusion occurs to a nerve without an abscess, one should pour hot, nonastringent olive oil over it time and again.

(٨٢) فصل: العصب إذا نخس حدث من ذلك وجع شديد يجلب إلى الموضع مادّة ‹فإن› يحصل عنها ورم في الموضع فينبغي أن يترك الخرق ويوسّع الجلد إن كان ضيّقا وأن يجتهد في تسكين الوجع بالتمريخ والتضمّد بالأدهان وبالأشياء الحارّة اللطيفة التي لا قبض فيها التي قد أسخنت ويحذر التنطيل والاستحمام بالماء الحارّ لفرط ترطيبه وذلك أنّ مادّة العصب رطبة قد جمّدتها البرودة فإذا رطّب وسخّن أسرع إليه العفن.

(٨٣) فصل (fol. 122b): الأدوية التي توافق قروح العصب هي الفاترة الحرارة اللطيفة الأجزاء التي لا قبض فيها ولا لذع كالزيت العذب وعلك البطم والحلتيت والجواشير والفربيون وينبغي أن تختار من هذا أضعف قوة للأبدان الرطبة ولمّا كان العصب مكشوفا تصل إليه قوة الدواء وما كان منها أشدّ قوة للأبدان اليابسة ولمّا كان من الأعصاب غائرا ألّا تصل إليه قوة الدواء وذلك أنّ قروح الأعصاب تحتاج أن تصل إليه من الأدوية حرارة فاترة وتجفّف قوي مع جوهر لطيف فإن كانت النخسة في اليد فامرخ الرقبة والإبطين بالدهن الحارّ المسخّن وإن كانت في الساق فامرخ الأربيتين. وينبغي أن تجعل أدوية قروح الأعصاب كلّها مسخّنة بالفعل لئلا يجلب بردها التشنّج ويصبّ في الخراج دهنا لئلا يزيد في وضره وإذا عرض للعصب رضّ من غير خراج ينبغي أن يصبّ عليه زيتا مسخّنا لا قبض فيه مرّة بعد أخرى.

---

٧ الحرارة] .emend. eds الحارارة MS || ٩ والفربيون] .emend. eds والفرابيون MS || ١٣ والإبطين] .emend. eds والإباطين MS || ١٤ الساق] .emend. eds السق MS الرجل MS²

(84) The setting of a broken bone is effected through straightening the limb [so that] the deflected part is brought into one line with the straight part above it. Then it should be put together by gently and carefully bringing in the splintered parts, one upon the other, and the fissure should be mended so that [the bone] returns to its initial shape. Then it should be bandaged, and splints should be put on it such that it reassumes it original shape. The splints should have the same shape as the [broken] limb so that they preserve [the shape]. To stop the bleeding, the bandage should be tighter on the site of the fracture and looser on the area around it; bind it high up on those sides. The broader the bandage, the better it is, for then it occupies a large area.

(85) In the beginning of the treatment of someone suffering from a fracture, nutrition should consist of fine foodstuffs, until the patient is out of danger. Then it should produce good, strong blood. Foodstuffs that produce this [kind of] blood are those that have good chyme and that are very nutritious, so that [the secreted nutrition] going [to the fracture] is glutinous and cohesive and the [parts of the] bone are linked [to one another] more firmly. A salve that helps the production of the matter exuding [from the wound] is that which is viscous and slightly heating.

(86) One should set one's mind on all the things discussed in this treatise, compare them with one another, and then act accordingly.[74] Beware of fear and of not being generous in [any matter, whether] small or large, and do not be excessive in something that the contemporary physicians consider to be needless. Do not treat evil diseases [so that] you will not be called a "physician of evil."[75] Do not deviate from the course pursued by physicians [in general], and do not try a medicine that has not been tried before [and found to be] safe.[76] I have mentioned all the precautions one should take in this regard in the treatise *On Purgatives*.[77] This is what I intended [to write], praise be to God. The copying was completed, abundant praise be to God, on October 3, [1424].

(٨٤) فصل: جبر العظم المنكسر يكون بتمديد العضو و ردّ الجزء المائل إلى موازاة الجزء القويم الذي فوقه واتّصاله وتلطّفه في دخول شظاياه بعضها في بعض بإحكام ويصلح ما فيه من الخلل إلى أن يرجع إلى شكله الأوّل ثمّ يلفّ عليه من اللفائف ما يمكن أن تصيّره مع الجبائر على شكله الأوّل. وينبغي أن تكون الجبائر على شكل ٥ العضو لتحفظ عليه شكله وأن يكون الرباط في موضع الكسر أشدّ و في حوالاه أرخى ليعصر الدم ويرفعه إلى تلك الناحيتين وكلّما كان الرباط أعرض كان أجود لأنّه يأخذ موضعا كبيرا.

(٨٥) فصل: الغذاء في أوّل علاج المكسور ينبغي أن يكون لطيفا (١٢٣a) إلى أن يؤمن المريض ومن بعد ذلك ينبغي أن يكون مولّد للدم المتين المحمود. والأغذية التي ١٠ تولّد هذا الدم هي المحمودة الكيموس الكثيرة الغذاء ليكون الرشح علكا لا ينقصف ليربط العظم رباطا أوثق واللطوخ الذي يعين في تولّد الرشح هو ما كان فيه قليل إسخان وفيه لزوجة.

(٨٦) فصل: ينبغي أن تجعل ذهنك في جميع الأشياء المتعلّقة في هذا الكتاب وقايس بينها ثمّ اعمل بحسبها ما ينبغي واحذر أن تخوّف ولا أن تسامح في صغير أو كبير ١٥ ولا تفرط في شيء لم يغنوه(؟) أطبّاء زمانك ولا تعالج علّة سوء ولا تسمّى طبيب سوء ولا تخرج عن الطريق الذي يقصده الأطبّاء ولا تجرّب دواء ما لم يكون مجرّبا مأمونا. وقد ذكرت جميع ما يحتاج في هذا التحرّز في كتاب الأدوية المسهلة. وهاهنا بلغ الغرض المقصود والحمد لله. تمّ الكتاب والحمد لله كثيرا يوم الثلاثة (ج) آقتوبر عام.

---

٢ الجزء] .emend. eds الجبر MS | وتلطّفه] .emend. eds للتلطفه؟ MS || ٥ حوالاه] .emend. eds حواله MS || ٩ المريض] .emend. eds المرض MS || ١٠ الرشح] MS[1] || ١٤ تخوّف] .emend. eds تجوّف MS || ١٥ أطبّاء] MS[1]

Supplement

On Rules Regarding the Practical Part of the Medical Art

3. Pay attention every day to improving the air that reaches the body through inhalation so that it will be perfectly balanced and free from all that might pollute it. The finer the pneuma is, the more sensitive it is to alterations in the air. The natural pneuma is coarser than the vital pneuma, while the vital [pneuma] is coarser than the psychical [pneuma].

15. The eye, because of its extreme sensitivity, will tolerate only [medications] that alleviate the pain, such as egg white and mucilage of fenugreek [*Trigonella foenum graecum*] and so also tolerates only those remedies that are extremely fat.

Other Medical Works by Maimonides

On Asthma 13.2–3; *On the Regimen of Health* 4.1
Galen said: Pay attention to the matter of the essence of the air which reaches the body through inhalation so that it will be utterly balanced and free from all that might pollute it. Says the author: The finer the pneuma is, the more it changes with changes in the air. The natural pneuma is coarser than the vital pneuma, while the vital pneuma is coarser than the psychical pneuma.

Medical Aphorisms 15.28
The eye is the most sensitive of organs. Therefore, drip medications into it after extremely gently lifting the upper eyelid. The medications should be steeped in a liquid that does not bite. The ancient [physicians] were very much in the right to use egg white [as a base for eyedrops]. *De methodo [medendi]* 13.

Medical Aphorisms 9.22
[For eye pain] one should apply to the eye a warm compress with a sponge [dipped] in water in which melilot and fenugreek have been cooked. . . . *Mayāmir* 4.

25. If [the pain] comes from inflating wind, it should be treated with foods, remedies, cataplasms, and fomentations that disperse the winds.

Medical Aphorisms 8.42
In the same manner one should persistently treat those pains that arise from an inflating wind with refining foods and drinks, enemas, cataplasms, fomentations and warm compresses. *De methodo* [*medendi*] 12.

27. If [someone's] stomach has a hot dyscrasia, he should be given cold water to drink. . . .

Medical Aphorisms 9.46
When a hot, bad temperament mixed with [either] a small amount of moisture [or dryness] dominates the stomach but does not affect the substance [of the stomach], we treat it with cold water. . . . *De methodo medendi* 7.

30. Cold water is appropriate for fever patients because it extinguishes a flare-up by cooling and moistening. [It is also appropriate for] severe anxiety and distress. It strengthens the intestines and dissolves biles in them, and it invigorates the body from [the weakness caused by] heat and dryness and prepares it for the intake of food. But one should not administer [cold water] when the noble organs are weak because of a tumor or because of a cold or moist dyscrasia. [And], by God, certainly not [when] the tumor is [...], and not before [the tumor] is ripe, nor during the crisis, in order to allow it to ripen. But if the body is well fleshed, dry heat dominates, the humors are concocted and may [already] have been evacuated, and there is no longer any weak organ within; then cold water should be given to drink.

Medical Aphorisms 10.2
Do not permit those suffering from fever to drink enough water to quench their thirst until you have thoroughly looked into the matter. When one of the noble organs is afflicted by an inflamed tumor [alone], or an inflamed tumor combined with erysipelas, or a soft or hard tumor; or when the body of the patient has an obstruction or a putrefaction of uncocted humors; or when it has an organ with a cold temperament, which is always harmed by cold water—in all these cases, the patient should not be given cold water to drink until there are clear signs that the putrefying humors are being cocted or that the inflamed tumor has become ripe. Someone suffering from a true erysipelas should be treated through the ingestion of cold water [under the same conditions]. *De methodo* [*medendi*] 9.

31. When [the fever] arises from cold, there is more need for hot air; but if it is accompanied by a catarrh or rheum and has not concocted, bathing should be postponed. . . .

Medical Aphorisms 19.35
If someone from those [who have] vaporous superfluities develops a fever from cold that befell him, he should take a bath, unless the fever is accompanied by a defluxion or catarrh. . . .
De methodo [medendi] 8.

35. Someone whose body is very emaciated or dominated by a dry dyscrasia should be given milk to drink. The best milk is that of women—after that, milk of a donkey, and then milk of a goat. The younger the animal, the better the milk, especially if it has been well fed and well provided for. The best milk to use is that which is sucked from the breast; and if that is impossible, one should bring the animal close to the patient so that he can [drink] the milk the moment it is milked, while it is still hot and has not cooled off. If one is afraid that it will curdle in his stomach, give [the patient] something with it that prevents [curdling] and that causes [the milk] to be quickly absorbed into the organs, such as honey and fresh milk. The purer the honey, the better it is, but sugar is [even] better.

Medical Aphorisms 20.39
Milk nourishes an emaciated body and revives it. It hinders the bad humors from causing harm and even improves them. It softens the stools. . . .
In Hippocratis Epidemiarum librum 2 *commentarius* 6.

Medical Aphorisms 21.12
Women's milk is the best of all milks for those suffering from marasmus, and after it comes the milk of donkeys.
De methodo medendi 7.

Medical Aphorisms 21.12
Milk has the property to change and alter very rapidly, just like sperm. Therefore, the best thing [to do], in the case of milk for someone who needs it, is to suck it from the breast.
De methodo medendi 7.

Medical Aphorisms 20.41
The milk which produces the best chyme is that which comes from an animal that is healthy and in a good condition, as long as one drinks it immediately after milking. One acts prudently if one adds a small amount of honey and salt to be sure that it does not turn into cheese.
De bonis [malisque] sucis.

37. When fevers originate from a surplus of blood, one should apply venesection until the patient is on the verge of fainting, if his strength allows it.

Medical Aphorisms 10.64
In all types of synochous fevers one should hasten to bleed the median cubital vein even to the point that the patient nearly faints, if his strength permits this. . . . *De methodo* [*medendi*] 9.

38. When the body of the patient is weak, when he suffers from indigestion, or when his body contains a crude humor that has not concocted, one should not bleed [that patient]. In the case of indigestion, one should postpone bleeding until the corruption of the food has diminished. Crude, viscous humors do not respond to expulsion through bleeding; and when one bleeds a body with weak organs, it becomes even weaker.

Medical Aphorisms 12.6
When the body is full with raw humors, it is very dangerous to apply venesection, because the strength [of the patient] is so greatly weakened and undermined that it is absolutely impossible for his body to return to its previous condition—and especially so if the patient also has fever.
De venae sectione.

Medical Aphorisms 12.11
If bleeding is required but indigestion occurs, one should postpone the bleeding until the food is digested and its superfluities are excreted from the body. . . . *De methodo* [*medendi*] 9.

39. If there is a surplus of blood in the body, it should be evacuated all at once, if possible. But if this is not possible, [it should be done] in several steps. If [the blood] is bad in quality, it should be evacuated little by little. Each [quantity] that is evacuated from the body will be replaced with wholesome blood.

Medical Aphorisms 12.15
When the body contains a large quantity of blood that has become extremely hot and seething and causes acute fever, one should evacuate a large quantity of blood at one stroke to the point of syncope, . . .
De venae sectione.

Medical Aphorisms 12.16
If someone whose strength is weak needs the evacuation of a large quantity of blood, the best thing is to do so in many sessions either in a single day or on a second and third day [as well]. . . . *De venae sectione.*

40. In all kinds of treatment, one should pay attention to the condition of the air, the external condition of the body, and the age [of the patient]. Thinness and a hot temperament dissolve the body; and a thin body that is affected by much dissolution, such as the body of a child, does not endure bleeding. The same holds good for a body with weak organs, such as the body of an old man, especially since an old man has a cold temperament and a small quantity of blood. If bile is produced in the cardia of the stomach of a body, [the person] is quick to vomit [and to develop] headache and syncope.

Medical Aphorisms 12.3
Do not bleed children under fourteen years, nor [anyone] older than seventy. Do not consider the number of years only, but also the external condition [of the body], because some people are [only] sixty years old and yet cannot tolerate venesection, while others who are seventy can tolerate it because they have much blood and their strength is great. But in spite of this, you should only extract a small amount of blood, even if their blood is like that of those who are in the prime of their life. *De venae sectione.*

Medical Aphorisms 9.43
I also saw patients with fever suffering from sudden spasms because of a bad humor that streamed to the cardia of the stomach and irritated it. When they vomited, they found immediate relief. . . . When the cardia of the stomach is affected, a sudden and rapid syncope may also arise from it. *De locis affectis* 5.

Medical Aphorisms 17.39
If someone complains of constant headache because of hypersensitivity of the nerves which spread in the cardia of the stomach, its treatment is a matter of the art of the regimen of health. . . . *De sanitate tuenda* 6.

42. If much blood has flowed to one of the organs and [the organ has] become occluded, the patient should be bled as soon as his strength and the other [conditions] are conducive.

Medical Aphorisms 12.14
It often happens that blood, before it putrefies, suddenly streams to one of the organs because of its surplus and either mortifies it completely, so that its function is annulled, or causes it severe damage. . . . For this disease originates from a large quantity of blood streaming to the brain. Therefore, when signs of a surplus of blood become evident, along with strength of the faculties, . . . carry out venesection without any caution. *De venae sectione.*

47. A hard tumor should be treated [first] by a softener, then by a discutient, and finally by a solvent. In general, one should see to it that the fine humors are not expelled and that the coarse humors do not remain, turning hard as stone and no longer reacting [to the solvents].

Medical Aphorisms 9.120
[For the treatment of] hard tumors, softening remedies should always be mixed with a discutient agent. . . .
De methodo [medendi] 14.

48. Scrofula affecting flesh that serves as filler (such as the flesh of the neck, the armpits, and the groin) should be treated by solvents alone, whereas scrofula affecting flesh that serves some useful purpose (such as the flesh of the breasts and [the flesh] that produces sputum) should be treated at first with repelling and strengthening [remedies], and then, when they reach their climax, with solvents. In general, all scrofula that do not react to solvents should be left to putrefy and then [should be] extirpated through surgery. During surgery one should be careful not to injure a nerve, artery, or vein.

Medical Aphorisms 15.19
Scrofula is a hard tumor that arises in the soft flesh. If it arises in the soft flesh that was created for an important function—namely, that which was created for the production of the sputum and the like—and pulsatile and nonpulsatile vessels are connected to it, its therapy should be similar to that of other hard tumors. But [the scrofula] arising in the soft flesh created to fill up empty space and to support the vessels should be treated through extirpation of the bad organ in its entirety. This [can be done] either through surgery, as is done in the case of cancer, or by letting it putrefy.
De methodo [medendi] 14.

52. If one applies surgery, one should examine [the cancer] minutely and excise it completely so that no trace of it is left. One should let the blood flow and press and squeeze the surrounding area in order to expel the hot, thick, melancholic blood from it. One should not be quick to stop [the flowing of the blood]. Afterward, one should treat the wound appropriately.

Medical Aphorisms 15.18
If you choose to treat a cancer through surgery, begin by evacuating the melancholic humor through purgation. Then excise the entire site until not even the root thereof remains. Let the blood flow, and do not hasten to stop it. Then compress the surrounding vessels and press the thick blood out of them. Treat [it in the same way as other] ulcers. *De methodo* [*medendi*] 14.

73. Diverting [surplus] matter to the opposite side is accomplished by bleeding from the opposite side, extracting the blood therefrom many times, and apply cupping glasses. In the same manner, if we see a nosebleed that increases, becoming more severe, which cannot be stopped with remedies, we bleed from the veins of the arm that is on the same side as the nostril from which the blood flows, and we apply cupping glasses to the hypochondria. If the blood flows from the right nostril, [we apply cupping glasses] to the liver; and if it flows from the left nostril, [we apply them] to the spleen so that the blood is attracted downward. We also tie the hands and feet [with bandages] once they have been rubbed.

Medical Aphorisms 9.2
In the case of a nosebleed you should not wait for the strength [of the body] to collapse. Rather, if you see the blood streaming rapidly, you should bleed the vein at the inner side of the arm on the same side as the hemorrhage, then you should tie the extremities with linen bandages, and then you should apply a cupping glass to the hypochondria on the side opposite to the nostril from which the blood flows. . . . *De venae sectione*.

Glossary of Technical Terms and Materia Medica

Guidelines for the Glossary

The following paragraphs describe the arrangement of entries and explain the use of symbols:

Arabic Entries

1. Order of entries: The glossary is arranged according to the Arabic roots. Within each root the following order has been applied: Verbs are listed first, followed by the derivative nominal forms in order of their length and complexity, then followed by the verbal nouns (*maṣdar*) of the derived stems and followed finally by the participles, both in the order of their verbal stems.

2. Verbs: Verbs are listed according to the common order of the verbal stems (I, II, III . . .). If the first stem does not appear in the text, the first derived stem to do so is introduced by the first stem, set in parentheses. Where more complex expressions headed by a verb are listed, they directly follow the corresponding verb.

3. Nouns: The different numbers of a noun (*sg.*, *du.*, *pl.*, *coll.*, *n. un.*) are listed as separate entries and are usually given in their indeterminate state. In a few cases, however, words are listed with the article instead. This practice is applied when the word is commonly used with the article in general or, if it always appears in the text, with the article in a nominalized usage.

4. Complex expressions: Each entry may have subordinate entries featuring complex expressions that contain the term from the superordinate

entry. Complex expressions may be listed in the indeterminate as well as in the determinate state.

5. Foreign words: Foreign words are listed in a strictly alphabetical order unless they are arabicized.

6. Vocalization: Only such words as might be confused with each other are vocalized. For the most part this applies to the verbal nouns of the first stem that might be confused with the verb. In these cases only the verbal noun is vocalized. Nouns that are distinguishable from each other by their vowel structure are likewise vocalized unless only one of them appears in the glossary.

7. Numbers: The numbers indicate the chapter and section of the Arabic text in which the respective entry may be found.

Use of Symbols in the Arabic Entries

1. – The en dash is used in subentries to represent the superordinate entry. If this superordinate entry is a complex one, the dash represents only its first element.

2. : A word followed by a colon may have two functions: a singular with a colon introduces a plural or dual, when the corresponding singular does not figure in the text. Any word followed by a colon may be used to introduce complex verbal or nominal expressions containing the word preceding the colon when this word itself does not figure in the text as an isolated item. The two functions of the colon may be combined.

3. : – An en dash followed by a colon introduces a complex entry that contains the superordinate word in a grammatically modified form.

4. ← The arrow, meaning "see" or "see also," refers to other entries either containing the word in question or representing a different orthography thereof.

English Equivalents

The English translation corresponds to the Arabic entry as it is translated in the English text. Therefore, it does not necessarily represent the common usage of the Arabic word independent from the text. This practice also means that there may be a lack of symmetry between the different translations of the singular, dual, and plural of a given word.

Number and determination are translated schematically in the glossary even if they are translated differently in the text. Therefore the translation of an Arabic term by itself does not always have to correspond grammatically to the translation required by the text.

Arabic verbal nouns (*maṣdar*) are never translated as English infinitives, in order to set them apart from Arabic finite verbs. Instead they are translated by any nominal form used as a corresponding translation in the English text. If the English text uses only finite verbs for a particular instance of an Arabic verbal noun, the glossary gives the gerund form (-ing).

Translation	Sec. No.	Maimonides	
		إبريسم ← خيط، شدّ	
bathing basin	34, 35	أبزن	1
		← دخل، دخول	
armpits	83	إبط: إبطان	2
		← لحم	
		أتن ← لبن	
trace	52	أثر	3
pears	33	إجّاص	4
to prepare	63	(أخذ) اتّخذ	5
		متّخذ ← حساء	
ear	65	أذن	6
		أذان ← قروح	
groin	23	أربية	7
		← لحم	
groins	83	أربيتان	8
sweet smells	26	أريج: أرائج	9
spinach	33	إسفاناخ	10
sponge	50, 79	إسفنجة	11
ceruse	62	إسفيداج	12
to extirpate	48	(أصل) استأصل	13
root of Florentine iris	62	أَصْل: أصل السوسن	14
extirpation	58	استئصال	15
opium	69	أفيون	16
acacia	63	أقاقيا	17

Translation	Sec. No.	Maimonides	
cadmia	62	إقليميا	18
to eat	31	أكل	19
eating	64	أَكل	20
cankering	56	تأكُل	21
		أكل ← أدوية	
to hurt	17	ألم	22
pain	80	أَلَم	23
safe	9, 76, 86	مأمون	24
sarcocol	62	أنزروت	25
		أنف ← قروح	
myrtle	63, 66	آس ← حبّ	26
the digestive organs, the respiratory organs	20, 45, 63, 64, 77	آلة: آلات الغذاء، آلات التنفّس	27
hiera picra	19	إيارج: إيارج الفيقرا	28
		باسليق ← فصد	
blood bursting forth, hemorrhage	56, 75, 76	انبثاق: انبثاق الدم ← أدوية	29
crisis	30	بحران	30
vapor	13, 70	بخار	31
smoky vapor(s)	11, 13, 31, 34	البخار: البخار الدخاني	32
moist, sweet vapor	34	ـ الرطب العذب	33
pleasant vapors	11	بخارات: البخارات العذبة	34

Translation	Sec. No.	Maimonides	
body	2, 3, 4, 5, 6, 7, 11, 16, 19, 20, 23, 24, 26, 27, 30, 31, 32, 35, 36, 38, 39, 40, 45, 51, 59, 71	بدن ← ترطيب، سحنة، تسخيف، استفراغ، تكثيف، إنضاج، تنقية	35
the body of an old man	40	ـ الشيخ	36
the body of a child	40	ـ الصبي	37
a moist body	58	البدن: البدن الرطب	38
a body with weak organs	38, 40	ـ الضعيف الأعضاء	39
a strong . . . weak body	58	ـ القوي. . . الضعيف	40
a bilious body	11	ـ المراري	41
a body that is extremely fat	12	ـ المفرط السمن	42
a body that is extremely lean	12	ـ المفرط الهزال	43
a thin body that is affected by much dissolution	40	ـ النحيف الكثير التحلّل	44
a dry body	13, 58	ـ اليابس	45
bodies	1, 8, 11, 13, 24, 34	أبدان	46
obese bodies	13	الأبدان: الأبدان الخصيبة	47
moist bodies	83	ـ الرطبة	48
bodies of a moderate [temperament]	23	ـ المعتدلة	49
lean bodies	13	ـ النحيفة	50
dry bodies	23, 83	ـ اليابسة	51
to be cured	77	برئ	52
to heal	75	تبرّأ	53

Translation	Sec. No.	Maimonides	
healing, cure	57, 60, 67, 77	برؤ	54
to cool off, to become cold	35, 79, 80	برد	55
to cool	11, 15, 27	برّد	56
cold	10, 23, 30, 31, 60, 70, 83	بَرْد	57
coldness of the spot	60	ـ الموضع	58
cold	17, 82	برودة	59
cooling	17, 18, 28, 33, 46	تبريد	60
cold	13, 28, 40, 60, 66	بارد ← سوء، فاكهة، ماء، هواء	61
cooled	27	مبرّد ← لبن	62
to protrude	79, 80	برز	63
excrements, excretion	6, 17	براز	64
seeds that expel the winds	32	بزر: بزور طاردة للرياح	65
coral	69	بسد	66
simple	58	بسيط	67
to make an incision	49	بطّ	68
to counteract	18	(بطل) أبطل	69
belly, stomach, abdomen	27, 32, 80	بطن ← مراق	70
the lower [part of the] belly	79	البطن: البطن الأسفل	71
		بطون ← طلق	
		باطن ← أعضاء، قروح	

Translation	Sec. No.	Maimonides	
to be put into action, to flow	5	(بعث) انبعث	72
it bleeds	75	انبعث منه دم	73
		بعيد ← أعضاء	
		باقلى ← دقيق	
remnant	16	بقية	74
to be moistened	23	(بلّ) انبلّ	75
a hot country	58	بلد: البلد الحارّ	76
puberty	10	بلوغ	77
phlegm	50	بلغم	78
urine, urination, micturition	6, 17, 36	بول ← دواء	79
room	34	بيت	80
whiteness	60	بياض	81
egg white	15	- البيض	82
		بيض ← بياض، صفرة	
indigestion	11	تَخم	83
indigestion	31, 38	تخمة	84
indigestions	32	تخم	85
theriac	69, 70	ترياق	86
exertion	11	تعب	87
apple	63, 69	تفّاح ← شراب	88
detrimental	20	متلف	89
tamarind	33	تمر: التمر الهندي	90

Translation	Sec. No.	Maimonides	
		توت ← ربّ	
tutty	62	توتيا	91
		ثدي: ثديان ← لبن،لحم	
omentum	79, 80	ثرب ← عروق	92
feces	80	ثفل	93
to weigh heavily	16	ثقل	94
snow	27	ثلج	95
		ثوب: ثياب ← لبس	
splints	84	جبيرة: جبائر	96
the setting of a broken bone	84	جبر: جبر العظم المنكسر	97
to curdle	35	(جبن) تجبّن	98
to irritate	31	(جحف) أجحف	99
		جدي ←لحم	
to draw, to attract, to pull	21, 45, 68, 73, 76	جذب	100
to have a strong attractive force	21	ـ اجتذابا قويا	101
to be drawn upwards	80	انجذب	102
to attract, to draw	45, 58	اجتذب	103
attraction	53	جَذْب	104
drawing the [superfluous] matter	21	ـ المادّة	105
the attraction of matter	46	اجتذاب: اجتذاب مادّة	106
elephantiasis	56	جذام	107
to try	86	جرّب	108

Translation	Sec. No.	Maimonides	
tried	86	مجرّب	109
a rattling sound	72	جرجرة	110
wound	75, 79, 81	جُرْح ← شفة، طرف	111
wound(s)	77, 78	جراح	112
wounds	77	جراحات	113
a pepper stomachic	32	جوارش: جوارش الفلافلي	114
lining, substance	19, 77	جرم	115
to flow	78	جرى	116
passages	15	مجرى: مجار	117
in a natural way	6	-: على المجاري الطبيعية	118
the straight part	84	جزء: الجزء القويم	119
the deflected part	84	الجزء المائل	120
hardness	43	جساء	121
to dry	11, 59, 62, 70, 72	جفّ: جفّف ← أدوية تجفيف ← أدوية	122
drying, dryness	56, 59, 65, 69	تجفُّف ← أدوية	123
intense dryness	83	- قوي	124
drying	27	مجفّف ← أدوية	125
to attract, to induce	32, 77, 82, 83	جلب	126
julep	72	جلاب	127

Translation	Sec. No.	Maimonides	
skin	76, 82	جلد	128
blossom of the wild pomegranate tree	62, 63	جلّنار	129
to cleanse	15, 33, 59, 62	جلا	130
cleansing, cleansing effect	19, 59	جلاء	131
to make solid	82	(جمد) جمّد	132
carbuncles	51	جمر	133
to accumulate	66	(جمع) اجتمع	134
		مجتمع ← تنقية	
opopanax	62, 83	جواشير	135
		جيّد ← دم	
		جوز ← ربّ	
hunger, empty stomach	11, 12	جوع	136
the chest cavity	77	تجويف: تجويف الصدر	137
substance	15	جوهر	138
the substance of the pneuma	26	- الروح	139
a fine substance	83	- لطيف	140
matters	1	جواهر	141
myrtle seed	72	حبّ: حبّ الآس	142
pomegranate seeds	23	- رمّان	143
to confine	32	حبس	144
diaphragm	77	حجاب	145
to turn hard as stone	47	(حجر) تحجّر	146

Translation	Sec. No.	Maimonides	
cupping	53	حجامة	147
cupping glasses	21, 42, 73	محجمة: محاجم ← تعليق	148
limit	46	حدّ	149
surgery	48, 58	حديد ← قطع	150
to go down, to descend	32	(حدر) انحدر	151
going down	35	انحدار	152
		حرّ ← أدوية	
heat	16, 17, 23, 30, 43, 46, 61	حرارة ← فاتر، ماء	153
the heat of the fever	18, 36	ـ الحمّى	154
moderate heat	83	ـ فاترة	155
the heat of the spot	60	ـ الموضع	156
the innate heat	2, 10, 14, 23, 26, 54	الحرارة: الحرارة الغريزية	157
hot, heat	11, 13, 23, 27, 35, 60, 66, 70, 79, 82	حارّ ← بلد، دم، دُهن، سوء، مزاج، ماء، نزلة، هواء	158
sharp	11, 13	حرّيف	159
		حركة ← رئة	
to feel	60	حسّ	160
sensitivity, senses	15, 25	حسّ	161
barley broth	11, 69	حسو: حسوالشعير	162
gruel, soup	23, 72	حساء	163

Translation	Sec. No.	Maimonides	
gruel prepared from wheat	23	ـ متّخذ من حنطة	164
astringent herbs	63	حشيشة: الحشائش القابضة	165
intestines	30	حشا: أحشاء	166
filler	48	حشو	167
		حصرم ← عصارة	
tightness	31	استحصاف	168
tight	26	مستحصف	169
lycium juice	62, 65	حضض	170
running	5	إحضار	171
decline of the fever	31, 33	انحطاط: انحطاط الحمّى	172
stretcher	34	محفّة	173
the preservation of health	1, 4, 14	حفظ: حفظ الصحّة	174
to give an enema (clyster)	32, 80	حقن	175
to become occluded, to be congested	42, 70	احتقن	176
clysters	64	(حقنة) حقن	177
obstruction	15	احتقان	178
		محكم ← خبز	
to dissolve, to untie the bandage	40, 75	حلّ	179
to dissolve, to clear	17, 23, 25, 31, 34, 43, 45, 47, 48, 70	حلّل	180
to dissolve	2	تحلّل	181
to dissolve	24, 31	انحلّ	182

Translation	Sec. No.	Maimonides	
		إحليل ← قروح	
dissolution, softening	24, 43, 45, 46, 47, 49	تحليل ← أدوية	183
dissolution	1, 45, 48, 51	تحلُّل ← بدن	184
dissolving	42	محلّل ← أدوية	185
to milk	35	حلب	186
		حلبة ← لعاب	
fresh milk	35	حليب	187
asafetida	83	حلتيت	188
to shave	70	حلق	189
		حَلْق ← إرخاء	
sweet	12, 15, 51	حلو ← أدوية	190
sweetness	15	حلاوة	191
to have fever	29	حمّ	192
fever	18, 20, 23, 28, 29, 31, 33, 45	حمّى ← حرارة، انحطاط، أعضاء	193
hectic fever	33, 34	– (ال)دقّ	194
ephemeral fever	29, 31, 32, 33	– يوم	195
fevers	30, 33, 37	حمّيات	196
continuous fevers	37	الحمّيات: الحمّيات الدائمة	197
fevers caused by repletion	36	– الامتلائية	198

Translation	Sec. No.	Maimonides	
bathing, bathhouse	12, 31, 34, 79	حمّام ← دخل، أدخل، إدخال	199
bathing	4, 7, 8, 11, 12, 23, 31, 82	استحمام	200
good	85	محمود ← خلط، دم	201
the red or black appearance	60	حمرة: حمرة اللون أو سواده	202
red	61	أحمر	203
redness	61	احمرار	204
sourness	31	حموضة	205
		حمل ← لحم	
applying suppositories	64	تحميل	206
to develop a fever	31	حمي	207
		حنطة ← حساء	
to be transformed	15	(حال) استحال	208
condition, case	1, 8, 16, 23, 31, 32, 40, 79	حال	209
conditions	1	أحوال	210
animal	35	حيوان ← نهشة	211
animals	35	حيوانات	212
to become malignant	56	(خبث) تخبّث	213
well-prepared bread	11, 16	خبز: الخبز المحكم الصنعة	214
sealing	58	خَتْم	215

Translation	Sec. No.	Maimonides	
being sealed	58	إختام	216
to benumb	25	(خدر) خدّر	217
carob	62	خروب	218
to be expelled, to leave	6, 32, 35, 38	خرج	219
to bleed	38	خرّج: خرّج الدم	220
to expel, to remove, to evacuate, to extract	4, 19, 23, 32, 36, 47, 49, 51, 71, 79, 80	أخرج	221
to apply venesection	37	ـ الدم	222
abscess	83	خراج	223
being expelled, leaving, exit, discharge	6, 31, 38, 53, 72	خروج	224
elimination, removal, expulsion, extraction	31, 49, 52, 53	إخراج	225
bleeding, extracting blood	68, 73	ـ الدم	226
defecating	80	ـ الزبل	227
		خردل ← دواء	
slenderness	24	انخراط	228
to be torn	77	(خرق) تخرّق	229
tear	82	خرق	230
a torn vein	68	انخراق: انخراق عرق	231
poppy	66	خشخاش	232
		خشن ← منديل	
		خشونة ← منديل	
to become well-fleshed	27	خصب	233

Translation	Sec. No.	Maimonides	
well-fleshed	30	خصيب ← أبدان	234
		اختصار ← فصل	
the testicles of roosters	11	خصية: خصى الديوك	235
to turn green	79	(خضر) اخضرّ	236
greenness	60	خضر	237
to stitch	81	خطّ	238
		خطر ← عرض	
		خفيف ← ضرب	
vinegar	43, 50, 62, 66	خلّ	239
vinegar mixed with a lot of water	72	- ممزوج بماء كثير	240
fissure	84	خلل	241
porous	26	متخلّل	242
to mix	17, 43, 44, 45, 63	خلط	243
humor	19, 20, 25, 33, 41, 47	خِلْط ← رقيق، غلظ	244
a humor mixed with blood	37	- مخالط للدم	245
a crude humor	38	- غليظ	246
thick, viscous humor	20, 25, 28	- غليظ لزج	247
raw humor	23	- فجّ	248
a biting humor	25	- لذّاع	249
good humors	16	الخلط: الخلط المحمود	250
the putrid humor	36	- العفن	251

Translation	Sec. No.	Maimonides	
humors	23, 24, 30, 37, 51	أخلاط ← رداءة، تعديل، تغليظ، تلطيف، إنضاج	252
the bad humors	71	الأخلاط: الأخلاط الرديئة	253
crude, viscous humors	38	ـ الغليظة اللزجة	254
delirium	77	اختلاط: اختلاط العقل	255
		مخالط ← خِلْط	
difference	15, 17, 63	اختلاف	256
disposition	17	خلق	257
scrofula	48	خنزير: خنازير	258
a silken thread	76	خيط: خيط إبريسم ← شدّ، طرفان، عقد	259
threads	81	خيوط	260
suturing	81	خياطة	261
to follow this regimen	34	دبر: دبّر بهذا التدبير	262
anus	76	دبر	263
regimen	52	تدبير ← دبر	264
		دجاج ← صفرة	
to go into the bathing basin	35	دخل: دخل الأبزن	265
to enter the bathhouse	32, 33	ـ الحمّام	266
to make enter the bathhouse	34	أدخل: أدخل الحمّام	267
the entry points of the veins	52	مدخل: مداخل للعروق	268
bringing in	84	دخول	269

Translation	Sec. No.	Maimonides	
going into the bathing basin	31	ـ الأبزن	270
taking to the bathhouse	33, 70	إدخال: إدخال الحمّام	271
		دخاني ← بخار، فضول	
smokiness	31	دخانية	272
		مدرّ ← دواء	
fat	12	دسم	273
to repel	48	دفع	274
repelling	51	دَفْع	275
cause	19	دافع	276
		دقّ ← حمّى	
bean flour and barley flour	71	دقيق: دقيق الباقلى والشعير ← شدّ	277
meal of bitter vetch	54, 62	ـ الكرسنّة	278
pounded	55	مدقوق	279
to rub	23, 68	دلك	280
massage	4, 7, 8, 11, 22, 23, 31	دَلْك	281
being rubbed	73	تدلُّك	282
blood	39, 42, 51, 52, 76, 78, 85	دم ← انبثاق، انبعث، خرّج، أخرج، إخراج، خِلْط، رداءة، عصر، قليل، كثرة، نزف	283
dragon's blood	62	ـ الأخوين	284
wholesome blood	14, 39	ـ جيّد	285
hot, melancholic blood	52	ـ حارّ سوداوي	286

Translation	Sec. No.	Maimonides	
coagulated blood	72	الدم: الدم المنعقد	287
thick, hot, melancholic blood	52	ـ الغليظ الحارّ السوداوي	288
good, strong blood	85	ـ المتين المحمود	289
brain	4, 63	دماغ	290
to make [something] form a scar	59	(دمل) أدمل	291
forming a scar	58	إدمال	292
forming a scar	58	اندمال	293
		دامل ← أدوية	
to pollute	3	دنس: دنّس	294
oil	31, 70, 83	دُهْن	295
hot (olive) oil	31, 79	ـ مسخّن	296
quince oil	68	ـ السفرجل	297
lukewarm oil	34	ـ مفتّر	298
nard oil	68	ـ الناردين	299
rose oil	50, 54, 62, 68	ـ (ال)ورد	300
hot oil	83	الدهن: الدهن الحارّ المسخّن	301
oils	63, 82	أدهان	302
rubbing	23	دَهْن	303
roundness	57	استدارة	304
constant	1	دائم ← حمّيات	305
to treat	52	(دوى) داوى	306
remedy, medicine	15, 67, 75, 83, 86	دواء	307

Translation	Sec. No.	Maimonides	
the mustard remedy	70	ـ الخردل	308
a medicine that stops the bleeding	75	ـ قاطع للنزف	309
a concocting medicine	51	ـ منضج	310
remedies, medicines, medication, drugs	15, 17, 25, 36, 43, 44, 45, 58, 60, 63, 64, 67, 69, 70, 71, 73, 83	أدوية ← قوى، كتاب	311
drugs that are more moist	58	ـ أرطب	312
remedies that have a strong drying effect	51	ـ شديدة التجفيف	313
remedies for ulcers	62	ـ القروح	314
medicines for all tumors in the nerves	83	ـ قروح الأعصاب كلّها	315
medicines that are less heat-producing	58	ـ أقلّ إسخانا	316
caustic remedies with a strong drying effect	65	ـ قوية التجفّف والكاوية	317
strong, restraining medicines	68	ـ قوية رديعة	318
strengthening remedies	43	ـ مقوّية	319
remedies that are more drying	65	ـ أكثر تجفّفا	320
drugs that eat away	58	الأدوية: الأدوية الأكلة	321
remedies that dry but do not burn	52	ـ التي تجفّف من غير تلذيع	322
drying medicines	69	ـ المجفّفة	323
solvents	46	ـ المحلّلة	324
sweet, strong medicines whose parts are fine and which dispel the winds	50	ـ الحلوة القوية اللطيفة الأجزاء المفشّة للرياح	325

	Translation	Sec. No.	Maimonides
326	medicines that make an ulcer form a scar	59	ـ الداملة
327	mineral [substances] that alleviate cough	63	ـ المعدنية المسكّنة للسعال
328	harsh (drastic) remedies	19	ـ العواصة(!)
329	remedies that evacuate [the residues] or that refine [the humors]	22	ـ التي تستفرغ والتي تلطّف
330	remedies that stop a hemorrhage	74	ـ القاطعة انبثاق الدم
331	strong remedies	17	ـ القوية
332	very hot drugs	61	ـ القوية الحرّ
333	remedies that are powerful solvents	49	ـ القوية التحليل
334	caustic medicines	76	ـ الكاوية
335	the medicines that cause an ulcer to close	59	ـ التي تلحم القروح
336	thinning remedies that do not heat excessively	25	ـ التي تلطّف من غير إسخان مفرط
337	refining drugs . . . diuretics	12	ـ الملطّفة. . .المدرّة للبول
338	emollient medicines	71	ـ الليّنة
339	medicines that make the flesh grow	59	ـ التي تنبت اللحم
340	vesicatory remedies	70	ـ المنفّطة
341	remedies appropriate for tumors in the nerves	83	ـ التي توافق قروح العصب
342	remedies that produce black bile	52	ـ المولّدة للكيموس السوداوي
343	treatment	64	مداواة
			ديك: ديوك ← خصية

Translation	Sec. No.	Maimonides	
diarrhea	31	ذرب	344
		← إفراط	
testicles	76	ذكر: مذاكير	345
mind	86	ذهن	346
to liquefy	55	ذاب	347
to dissolve	23, 30, 50	أذاب	348
		ذائب ← سمن	
the dissolution of the flesh	61	ذوبان: ذوبان اللحم	349
the dissolution of soft flesh	61	ـ اللحم الرخص	350
lung(s)	45, 64, 68, 70, 71, 72	رئة	351
		← قروح	
the lungs are organs that move constantly	67	ـ: الرئة عضو دائم الحركة	352
resin	62	راتينج	353
head	17, 31, 56, 70	رأس	354
major	4	رأيسي	355
a rob of mulberries	65	ربّ: ربّ التوت	356
a rob of walnuts	65	ـ الجوز	357
to apply a bandage, to bind	75, 76	ربط	358
to link the [parts of the] bone [to one another] more firmly	85	ـ العظم رباطا أوثق	359
ligament, bandage	55, 75, 76, 84	رباط	360
asthma	0	ربو	361
		ترتيب ← سوء	

Translation	Sec. No.	Maimonides	
feet	68, 80	رجل: رجلان ← شدّ	362
		رخص ← ذوبان	
		رخو ← لحم، ورم	
relaxing the muscles of the throat	71	إرخاء: إرخاء عضل الحلق	363
softening	45	مُرخ	364
the restoration of continuity	58	ردّ: ردّ الاتّصال	365
bad humors	22	رداءة: رداءة الأخلاط	366
badness of the blood, bad blood	60	ـ الدم	367
a bad treatment	60	ـ العلاج	368
		رديع ← أدوية	
to sprinkle	34	رشّ	369
the matter exuding [from the wound]	85	رشح	370
leaden [color]	60	رصاصية	371
contusion	83	رضّ	372
to moisten	10, 11, 34, 82	(رطب) رطّب	373
moist	13, 27, 28, 60, 66, 82	رطب ← بخار، بدن، أبدان، سوء، منديل	374
		أرطب ← أدوية	
moisture	34	رطوبة	375
a very thin and fine honeylike moisture	49	ـ أرقّ وألطف كالعسلية	376
a very thick moisture	49	ـ أغلظ	377

Translation	Sec. No.	Maimonides	
moistening [effect]	30, 33, 82	ترطيب	378
moistening of the body	7	ـ البدن	379
nosebleed	73	رعاف	380
		مرعوف ← منخر	
two bandages	57	رفادة: رفادتان	381
		مرتفع ← فراش	
softness, gentleness	7, 75, 80	رِفق	382
		رفيق ← هزّ	
thinness	24, 40, 77	رقّة	383
thin, fine	26, 47, 61, 72	رقيق ← شراب، كثر	384
the fine part of the humor	23	ـ الخلط	385
		أرقّ ← رطوبة	
hypochondria	73	مراقّ: مراقّ البطن	386
neck	83	رقبة	387
to be composite	27	(ركب) رُكِّبَ	388
to be composite	27	تركّب	389
knee	23	ركبة ← مفصل	390
composition	58	تركيب	391
composed	1	مركّب	392
		رماد ← ماء	
		رمّان ← حبّ، شراب، قشر، ماء	
swelling	61	رهل	393

Translation	Sec. No.	Maimonides	
		ترهّل ← شدّة	
to let rest	9, 23, 75	(راح) أراح	394
to rest, to find relief	35, 60	استراح	395
pneuma	3, 14, 54	روح ← جوهر	396
the natural pneuma	3	الروح: الروح الطبيعي	397
the vital [pneuma]	3	ـ الحيواني	398
the psychical [pneuma]	3	ـ النفساني	399
flatulence	50	ريح	400
inflating wind	25	ـ نافخة	401
winds	16, 25	رياح ← بزر، أدوية	402
winds	14	أراييح	403
exercise	4, 5, 7, 8, 9, 11, 12, 21, 22, 23	رياضة	404
to exercise	6	ـ: استعمل الرياضة	405
excessive exercise	11	الرياضة: الرياضة المفرطة	406
filtered	15	مروّق	407
rhubarb / currant	63	ريباس	408
sputum	48	ريق	409
vitriol	62	زاج	410
		زبل ← إخراج	
birthwort	62	زراوند	411
arsenic	76	زرنيخ	412

Translation	Sec. No.	Maimonides	
disturbance	18	إزعاج	413
saffron	54, 69	زعفران	414
pitch	12	زفت	415
rheum	31	زكام	416
pure	44	زكي	417
chronic	33	مزمن	418
verdigris	76	زنجار	419
angles	57	زاوية: زوايا	420
olive oil	32, 50	زيت	421
hot, nonastringent olive oil	83	ـ مسخّن لا قبض فيه	422
sweet olive oil	83	الزيت: الزيت العذب	423
to become more severe	73	زاد: زاد وفرط	424
increase	3, 12, 16, 61	زيادة	425
appendages	57	زائدة: زوائد	426
cause, reason	17, 18, 29, 31, 58, 59, 67	سبب	427
that which causes fainting	26	السبب: السبب الفاعل للغشي	428
things that cause	11	أسباب	429
the established factors [of health]	2	الأسباب: الأسباب المسنّة	430
the correctly proportioned causes	1	ـ المتناسبة	431
to be softened	23	(سحق) تسحّق	432
external condition	58	سحنة	433

Translation	Sec. No.	Maimonides	
the external condition of the body	40	ـ البدن	434
thinning the body	26	تسخيف: تسخيف البدن	435
to become warm	23	سخن	436
to heat, to warm	10, 32, 79, 82	سخّن	437
to heat	27, 82	أسخن	438
		سُخْن ← شراب	
hotter	34	أسخن	439
heat	61	سخونة	440
heating	17, 46	تسخين	441
heating	7, 33, 34, 45, 59, 63, 85	إسخان ← أدوية	442
hot, calefacient	83	مسخِّن ← دُهْن، زيت	443
to close	11	سدّ	444
rue	32, 50	سذاب	445
pleasure	29	سرور	446
cancer	52	سرطان	447
quickly digested	69	سريع: سريع الانهضام	448
the forearms	23	ساعد: الساعدان	449
		سعال ← أدوية، سكون، إسكان، تشاغل	
quince	63, 69, 72	سفرجل ← دُهْن	450
		أسفل: سفلى ← أعضاء	

Translation	Sec. No.	Maimonides	
a collapse of strength	18	سقوط: سقوط القوة	451
to let drink	16, 27, 30, 32, 35, 72	سقى	452
giving to drink	35	سَقْي	453
those suffering from dropsy	50	مستسقٍ: مستسقون	454
sugar	35	سكر	455
to subside, to be stopped	23, 73	سكن	456
to alleviate, to stop	15, 25, 76, 80	سكّن	457
rest	68	سكون	458
alleviating the cough	71	ـ السعال	459
houses	23	مسكن: مساكن	460
alleviating the pain	82	تسكين: تسكين الوجع	461
alleviating the pain	26	ـ الأوجاع	462
alleviating the cough	69	إسكان: إسكان السعال	463
		مسكّن ← أدوية	
oxymel	23, 33	سكنجبين	464
those suffering from phthisis	50	مسلول: مسلولون	465
atheromas	49	سلعة: سلع	466
		سلق ← أطراف	
boiled	55	مسلوق	467
the soundness of the organs	14	سلامة: سلامة الأعضاء	468
poison	58	سمّ	469
		سموم ← وهج	

Translation	Sec. No.	Maimonides	
pores	11, 24, 26	مسامّ	470
		مسموم ← نهشة	
sumac	23, 63, 65	سمّاق	471
fishes	16, 69	سمك	472
rockfish	11	السمك: السمك الرضراضي	473
to make more fat	12	(سمن) أسمن	474
fat	12, 15	سمن	475
old butter	55	- عتيق	476
melted butter	55	السمن: السمن الذائب	477
age	40	سنّ	478
Indian nard	69	سنبل	479
sleeplessness	11, 29	سهر	480
to be easy	67, 72	سهل	481
to ease	72	سهّل	482
purgatives	25	مسهل: مسهلات ← كأب	483
facilitation	53	تسهيل	484
laxation, purgation	19, 23, 36, 71	إسهال	485
		سوء ← طبيب، علّة	
their bad arrangement	60	- ترتيبها	486
bad temperament	19, 20, 59	- مزاج	487
a cold dyscrasia	27	- مزاج بارد	488
a cold or moist dyscrasia	30	- مزاج بارد أو رطب	489

Translation	Sec. No.	Maimonides	
a hot dyscrasia	27	- مزاج حارّ	490
a hot, dry dyscrasia	28	- مزاج حارّ يابس	491
a moist dyscrasia	27	- مزاج رطب	492
a bad temperament of the flesh	60	- مزاج اللحم	493
a dry dyscrasia	27	- مزاج يابس	494
dyscrasia, bad temperament	27, 58	- المزاج	495
a dry dyscrasia	35	- المزاج اليابس	496
to turn black	79	(سود) اسودّ	497
		أسود ← شراب	
		سواد ← حمرة	
		سوداوي ← دم، أدوية	
		سوسن ← أصل	
leg	83	ساق	498
legs	23	ساقان	499
to flow	52, 78	سال	500
fluidity	1, 24	سيلان	501
alum	50, 62	شبّ	502
youth	35	شباب	503
a full stomach	12	شبع	504
to stop	75	شدّ	505
ligating it with a fine silken thread	79	شدّ: الشدّ بخيط دقيق من إبريسم	506
tying the hands and feet tightly	73	- اليدين والرجلين شدّا عنيفا	507
severity	18	شدّة	508

Translation	Sec. No.	Maimonides	
severe flabbiness	60	ـ الترهّل	509
intense dryness	60	ـ القحل	510
		شديد ← وجع	
to drink	35, 64	شرب	511
to soak	50	شرّب	512
wine	16, 51, 54, 63, 72	شراب	513
apple syrup and pomegranate syrup	15	ـ الرمّان والتفّاح	514
astringent, hot wine	79	ـ قابض مسخّن	515
astringent, diluted wine	23	ـ قابض ممزوج	516
chicory syrup	15	ـ الهندباء	517
thinning beverages	12	الشراب: الشراب الرقيق	518
hot wine	66	ـ السخن	519
astringent black wine	80	ـ الأسود القابض	520
old, pure wine	27	ـ العتيق الصرف	521
thickening beverages	12	ـ الغليظ	522
watery, astringent wine	16	ـ المائي القابض	523
drinking, taking	64, 71	شَرْب	524
drinking water	16	ـ الماء	525
beneath the ribs	32	شرسوف: ما تحت الشراسيف	526
scarification	21, 53	شَرْط	527
		شريف ← ضُعف، عضو، أعضاء، أفعال	
to share	17	(شرك) شارك	528

Translation	Sec. No.	Maimonides	
reciprocity	17	مشاركة	529
artery	56	شريان	530
arteries	48	شرايين	531
splintered parts	84	شظية: شظايا	532
		شعير ← حسو، دقيق، كشك، ماء	
refraining from coughing	68	تشاغل: التشاغل عن السعال	533
		شفاف ← غلظ	
lips	61	شفة: شفاه	534
the two edges of the wound	80	شفتان: شفتا الجرح	535
shape	84	شكل	536
		شكاية ← مقالة	
spasms	77, 83	تشنّج	537
bones	11	شوك	538
an old man	40	شيخ	539
		← بدن	
old people	10	مشائخ	540
to pour	83	صبّ	541
[to suffer from] a discharge, to flow, to be poured	9, 19, 23, 34, 42, 58	انصبّ	542
influx, streaming, stream, flowing, flow	19, 20, 23, 45, 50, 58	انصباب	543
the flow of matter	61	ـ مادّة	544
aloe	62, 65	صبر	545
		صبي ← بدن	
children	10	صبيان	546

Translation	Sec. No.	Maimonides	
health	1, 2, 8, 14, 20	صحة ← حفظ	547
sound, healthy [part]	20, 79	صحيح ← عصب	548
		صديد ← قلة، كثر	
chest	45, 68, 72	صدر ← تجويف	549
headache	40	صداع	550
wrestling	5	صراع	551
		صرف ← شراب	
		صعب ← علاج	
the yolk from chicken eggs	11	صفرة: صفرة بيض الدجاج	552
yellow bile [fever]	33	صفراوي	553
purer	35	أصفى	554
to become hard	59	صلب	555
hard / spinal column	7, 23, 59	صُلْب ← ورم	556
hardness, rigidness	43, 46, 61	صلابة	557
improving	51	إصلاح	558
the improvement of the water and [then the improvement of] foods	1	ـ الماء والأغذية	559
improving breathing	14	ـ التنفّس	560
the improvement of the air	1, 3	ـ الهواء	561
gum [arabic]	63	صمغ	562
hook	76	صنارة	563

Translation	Sec. No.	Maimonides	
		صنعة ← خبز	
art, function	5, 10	صناعة	564
		← كتاب	
piece of wool	79	صوفة	565
fasting	13	صوم	566
summer	58	صيف	567
harm	34	ضرر	568
beating lightly	12	ضرب: ضرب خفيف	569
to be weak(ened)	15, 23, 24, 67	ضعف	570
weakness	38	ضُعْف	571
weakness of a noble organ	28	- عضو شريف	572
the weakness of their vigor	24	- القوة	573
weak	16, 20, 30, 38, 58	ضعيف	574
		← بدن، عضو، أعضاء	
to block, to draw together, to bring into contact	15, 27, 80	ضمّ	575
to apply a poultice	27, 32, 68	(ضمد) ضمّد	576
poultice	44	ضماد	577
cataplasms, poultices	25, 44, 54	أضمدة	578
strong poultices	64	الأضمدة: الأضمدة القوية	579
poultices	45	ضمادات	580
plastering	82	تضمُّد	581
difficulty of breathing	72	ضيق: ضيق نفس	582
narrow	82	ضيّق	583

Translation	Sec. No.	Maimonides	
		طبّ ← كآب	
physician of evil	86	طبيب: طبيب سوء	584
physicians	2, 33, 86	أطبّاء	585
to cook	32, 66, 69, 72	طبخ	586
nature	33	طبع	587
nature	29, 57, 59, 77	طبيعة	588
		طبيعي ← مجرى، روح	
spleen	15, 56, 73	طحال	589
Maltese mushroom	62	طراثيث	590
		طارد ← بزر	
both edges	81	طرف: طرفان	591
the ends of the thread	79	طرفا الخيط	592
both edges of the wound	81	ـ الجرح	593
extremities	50	أطراف	594
offshoots of beets	55	ـ السلق	595
to feed	27	(طعم) أطعم	596
meal, food	12, 13, 16, 31, 32	طعام ← فساد	597
the putrefied food	31	الطعام: الطعام الفاسد	598
to extinguish	30	(طفئ) أطفأ	599
extinguishing	18	تطفية	600
extinction	36	إطفاء	601
to float	16	طفا	602

Translation	Sec. No.	Maimonides	
they suffer from diarrhea	23	(طلق) استطلقت بطونهم	603
to rub	12	طلى	604
Armenian clay, Armenian earth	62, 63, 69	طين: الطين الأرميني (الأرماني)	605
sealed earth	69	الطين المختوم	606
external	17	ظاهر ← أعضاء	607
		عتيق ← شراب	
to balance, to moderate	16, 58, 71	(عدل) عدّل	608
balancing, bringing into equilibrium, bringing into balance	3, 22, 29	تعديل	609
the balancing of the humors	46	ـ الأخلاط	610
balance, equality, moderation, normal condition	3, 27, 50, 71	اعتدال	611
moderate	11, 23	معتدل ← أبدان، ماء	612
mineral	52, 62	معدني ← أدوية	613
harm	25	عادية	614
		عذب ← بخار، زيت، ماء، مياه	
symptom(s), affliction, incident	18, 28, 50	عرض	615
severe symptoms	18	ـ خطر	616
symptoms	17	أعراض	617
to sweat	34	عرق	618
to make *s.o.* sweat	31	عرّق	619
sweat	34	عَرَق	620

Translation	Sec. No.	Maimonides	
vein	76	عِرق	
		← انخراق	
ischias	56	ـ النساء	621
(blood) vessels, veins	3, 6, 48, 77	عروق	622
		← مدخل	
the veins of the arm	73	ـ اليد	623
the blood vessels in the omentum	79	العروق: العروق التي في الثرب	624
promoting perspiration, sweating	12, 22	تعريق	625
to undress	34	(عري) عرّى	626
to be hard	60	عسر	627
honey	35, 65, 72	عسل	628
		← ماء	
		عسلي ← رطوبة	
nerve(s)	48, 55, 77, 82, 83	عصب	629
		← أدوية	
a healthy nerve	55	العصب: العصب الصحيح	630
nerves	77, 83	أعصاب	631
		← قروح	
		عصبي ← أعضاء	
to squeeze	52	عصر	632
to stop the bleeding	84	ـ الدم	633
juice of unripe, sour grapes	65	عصارة: عصارة الحصرم	634
the bite of a mad dog	58	عضّة: عضّة الكلب الكلب	635
the upper arms	23	عضد: العضدان	636

Translation	Sec. No.	Maimonides	
		عضل ← إرخاء	
part of the body, organ, limb	12, 17, 21, 42, 46, 53, 58, 59, 64, 80, 84	عضو ← رئة، ضُعف، تقوية، تمديد	637
a noble organ	23	ـ شريف	638
weak organ	30	ـ ضعيف	639
organs, bodily parts, limbs	2, 3, 4, 7, 9, 15, 16, 17, 19, 23, 24, 34, 35, 42, 43, 50, 54, 58, 63, 65	أعضاء ← بدن، سلامة، قوى، تقوية	640
the internal organs (parts)	17, 44, 66	الأعضاء: الأعضاء الباطنة	641
the remote organs	15, 44, 64	ـ البعيدة	642
the lower parts of the body	21	ـ السفلى	643
parts affected by high fever	17	ـ الشديدة الحمّى	644
the noble organs	23, 30, 43	ـ الشريفة	645
weak parts	17	ـ الضعيفة	646
the external parts	66	ـ الظاهرة	647
nervous parts	77	ـ العصبية	648
the upper parts	21	ـ العليا	649
the organs that are close	15	ـ القريبة	650
the fleshy parts	12	ـ اللحمية	651
aromatic	17	عطري	652
thirst	11, 16	عطش	653
to increase	12	(عظم) عظّم	654

Translation	Sec. No.	Maimonides	
bone	55	عَظْم	655
		← جبر، ربط	
magnitude	18	عِظَم	656
astringent	51, 72	عفص	657
gallnut(s)	62, 63	عفص	658
to putrefy	51	عفن	659
to let putrefy	48, 49, 55	عفّن	660
to putrefy	36	تعفّن	661
putrefaction, decay	11, 24, 32, 54, 76, 81, 82	عَفَن	662
putrefaction	20, 56, 65, 76	عفونة	663
		عَفِن ← خلط	
putrefaction	51	تعفُّن	664
to fasten the thread with a double knot	81	عقد: عقد الخيط عقدتين	665
		عقدة: عقدتان ← عقد	
tied up in the middle	80	متعقّد(؟): متعقّد(؟) الوسط	666
		منعقد ← دم	
		عقل ← اختلاط	
disease	56	علّة	667
evil diseases	86	ـ سوء	668
diseases	15	علل	669
patient	23, 35, 38, 42, 60, 68, 72, 75, 80	عليل	670

Translation	Sec. No.	Maimonides	
to treat	25, 47, 52, 86	عالج	671
treatment, therapy	1, 15, 17, 18, 22, 23, 27, 31, 32, 40, 41, 48, 49, 50, 58, 68, 76	علاج ← رداءة	672
the treatment of someone suffering from a fracture	85	ـ المكسور	673
a hard treatment	17	العلاج: العلاج الصعب	674
to apply	21	علق	675
to raise	80	علّق	676
the application of cupping glasses	45, 50, 73	تعليق: تعليق المحاجم	677
mastic from the turpentine tree	83	عِلْك: علك البطم	678
glutinous	85	عَلِك	679
indication	60	علامة	680
		أعلى: عليا ← أعضاء	
		أعلى: أعال ← فراش	
to use, to employ, to apply	35, 42, 51, 52, 53, 56, 64	عمل: استعمل ← رياضة	681
		عملي ← كتاب	
		عنق ← لحم	
habit	58	عادة	682
to prevent	0	عاق	683
fatigue	23	إعياء	684
		عيش ← قرار	

Translation	Sec. No.	Maimonides	
eye	15	عين ← قروح، نظر	685
to feed	31, 32	غذا	686
to feed	27	غذّى	687
to take food	24	اغتذى	688
food, nutrition	6, 13, 16, 20, 23, 24, 68, 85	غذاء ← آلة، تقليل، توسّع	689
foods, foodstuffs, nutrition	10, 11, 15, 19, 25, 26, 35, 36, 64, 69, 85	أغذية ← إصلاح، قوى	690
soft foods	71	الأغذية: الأغذية الليّنة	691
nutrition	14, 24	تغذية	692
intake of food	30	اغتذاء	693
		غريزي ← حرارة	
to gargle	64	(غرغر) تغرغر	694
to agglutinate, to be agglutinant	64, 74	(غرو) أغرى	695
to cleanse, to wash	25, 66	غسل	696
fainting, syncope	18, 26, 28, 37, 40	غشي ← سبب	697
anger	11, 31	غضب	698
to thicken	15, 24, 72	(غلظ) غلّظ	699
viscosity, coarseness, thickness	11, 33, 43	غلظ	700
the thickness of the humor	20	ـ الخلط	701
the opacity of the ulcer (lesion)	60	ـ شفاف القرحة	702

Translation	Sec. No.	Maimonides	
coarse, crude, thick	3, 23, 26, 47, 61	غليظ ← خلط، أخلاط، دم، شراب	703
coarser	3	أغلظ ← رطوبة	704
thickening the humors	26	تغليظ: تغليظ الأخلاط	705
boiled	50	مغلّى	706
to press	52	غمز	707
to immerse, to dip	31, 34, 50	غمس	708
to be immersed	34	انغمس	709
lying deep within	83	غائر	710
to deviate, to alter, to change	1, 17	(غير) تغيّر	711
alteration	3	تغيُّر	712
to open	15	فتح	713
when the attack abates	20	فتور: في وقت فتور النوبة	714
lukewarm	83	فاتر: فاتر الحرارة ← حرارة، ماء	715
		مفتّر ← دهن	
		فجّ ← خلط	
thighs	23	فخذ: الفخذان ← قروح	716
spurge	83	فربيون	717
		فرج ← قروح	
chickens	16	فرّوج: فراريج	718
chickens that lay eggs	11	فراريج: الفراريج البائضة	719

Translation	Sec. No.	Maimonides	
a flat bed raised at its extremities	80	فراش: فراش وطئ مرتفع الأعالي	720
		فرط ← زاد	
excess	82	فَرْط	721
excessive diarrhea	18	إفراط: إفراط الذرب	722
to evacuate	17, 30, 33, 39, 46	(فرغ) استفرغ ← أدوية	723
evacuation, emptying	16, 19, 21, 22, 25, 36, 42, 45	استفراغ ← تلطيف	724
the emptying of the body	46, 53	ـ البدن	725
a dissolution of continuity	68	تفرق: تفرّق الاتّصال	726
to rub	55	فرك	727
to degenerate	31, 55	فسد	728
degeneration	54, 55	فساد	729
the corruption of the food	38	ـ الطعام	730
degenerated	55	فاسد ← إخراج، طعام	731
causing to degenerate	53	مفسد	732
to disperse	25, 70	فشّ	733
		مفشّ ← أدوية	734
to bleed	38, 42, 73	فصد	735
bleeding, venesection	22, 23, 33, 36, 37, 38, 40, 41, 42, 73, 78	فَصْد	736
bleeding the basilic vein	68	ـ الباسليق	737

Translation	Sec. No.	Maimonides	
aphorisms	0	فصل: فصول	738
the seasons of the year	10	فصول: فصول السنة	739
in a concise, aphoristic way	0	-: على طريق الفصول والاختصار	740
joint	55	مفصل	741
the joint of the knee	23	ـ الركبة	742
joints	77	مفاصل	743
surplus	65	فَضْل	744
residues, superfluities	4, 6, 10, 14, 17, 71	فضول ← تنقية	745
the vaporous residues	10	الفضول: الفضول الدخانية	746
		فاطر ← مياه	
activity	4	فِعْل	747
an activity that is of major importance	4	ـ شريف	748
functions, activities	10, 17	أفعال	749
noble activities	17	ـ شريفة	750
fruit	27	فاكهة	751
cold fruits	31	الفواكه: الفواكه الباردة	752
agriculture	5	فلاحة	753
cardia, mouth	64, 71	فم ← قروح، مسك	754
the cardia of its stomach	40	ـ معدته	755
to annihilate	59	(فني) أفنى	756
opening	15	فوّهة	757

Translation	Sec. No.	Maimonides	
astringency	15, 69, 82, 83	قَبْض ← زيت	758
astringent	17, 24, 44, 45, 68, 74	قابض ← حشيشة، شراب، مياه	759
		قَحَل ← شدّة	
dry	66	قَحِل	760
to calculate, to evaluate, to determine	17, 18, 41, 64	(قدر) قدّر	761
size, amount, magnitude, quantity	12, 16, 17, 20, 50, 51	مقدار	762
according to, of the magnitude of, as much as	41, 46, 59	-: بمقدار	763
surplus	39	المقدار: المقدار الزائد	764
the right quantity	7	ـ القصد	765
the right quantity	7	-: القصد المقدار	766
assessment	2	تقدير	767
foot	23	قدم	768
a quiet life	14	قرار: قرار العيش	769
		قريب ← أعضاء	
wound, ulcer	52, 57, 58, 59, 61, 66, 67, 69, 70, 72	قرحة ← غلظ، توسيع، اتّصال	770
ulcers, tumors	57, 58, 59, 60, 63, 64, 66, 67	قروح ← أدوية	771
ulcers in the ears	65	ـ الأذان	772
ulcers in the nose	65	ـ الأنف	773

Translation	Sec. No.	Maimonides	
ulcers in the penis, vagina, and anus	65	ـ الإحليل والفرج والمقعدة	774
ulcers in the lungs	67, 69, 72	ـ الرئة	775
tumors of the nerves	83	ـ الأعصاب	776
tumors in the eyes	65	ـ العين	777
ulcers in the thighs	65	ـ الفخذين	778
ulcers in the mouth	65	ـ الفم	779
internal ulcers	69	القروح: القروح الباطنة	780
yellow amber pills	69	قرص: أقراص الكهرباء	781
burned papyrus	65	قرطاس: القرطاس المحروق	782
piece of cloth	34	مقرمة	783
pomegranate peels	62, 63, 66	قشر: قشور (ال)رمّان	784
scab	76	قشرة	785
scrape off	76	قشط	786
branches	12	قضيب: قضبان	787
to cut	80	قطّ(؟)	788
to dilute, to thin, to act as a discutient, to cut (off)	19, 33, 47, 55, 79, 81	قطع	789
to apply surgery	52	ـ بالحديد	790
to be cut off	57	انقطع	791
diluting, excising, cut, stopping	3, 52, 69, 79	قَطع	792
dissolving, a discutient effect	25, 43	تقطيع ← تلطيف	793
		مقعدة ← قروح	

Translation	Sec. No.	Maimonides	
		قفاء ← استلقاء	
small quantity	13	قلّة	794
lack of pus	60	ـ الصديد	795
speaking little	68	ـ الكلام	796
a small quantity of blood	40	قليل: قليل الدم	797
eating less food, diet	12, 22	تقليل: تقليل الغذاء	798
heart	4, 45	قلب	799
to remove, to undo	25, 75	قلع	800
rooting out	18	قَلْع	801
anxiety	30	قلق	802
green vitriol	76	قلقطار	803
roasted food	27	قلايا	804
		قوانين ← كتاب	
hollowing out	57	تقوير	805
the treatise on the [illness]	0	مقالة: المقالة في الشكاية	806
to counter	58	(قوم) قاوم	807
to strengthen	4, 10, 19, 24, 30, 31, 32, 48, 72	(قوي) قوّى	808
strength, capacity, power, vigor	15, 16, 17, 18, 20, 33, 36, 37, 41, 42, 44, 54, 58, 68, 79, 83	قوة ← سقوط، ضُعْف	809
powers, strength, potency	16, 17, 43, 67	قوى	810

Translation	Sec. No.	Maimonides	
the powers of the medicines and foods	58	ـ الأدوية والأغذية	811
the faculties of the organs, the powers of the bodily parts, the powers of the organs, the capacities of the parts	10, 16, 17, 26	ـ الأعضاء	812
		قوي ← بدن، جذب، تجفّف، أدوية، أدمضة	
strengthening of the organs	56	تقوية: تقوية العضو	813
strengthening of the organs	45	ـ الأعضاء	814
		مقوّ ← أدوية	
emesis, vomiting	17, 19, 32, 36, 40	قيء	815
to compare	86	(قيس) قايس	816
liver	3, 4, 6, 15, 17, 20, 23, 45, 52, 73	كبد	817
livers	24	أكباد	818
sulphur	76	كبريت	819
		كتب ← سلك	
the treatise *On Purgatives*	86	كتاب: كتاب الأدوية المسهلة	820
the treatise on rules regarding the practical part of the medical art	title	ـ قوانين الجزء العملي من صناعة الطبّ	821
a treatise on rules regarding the practical part of medicine	0	ـ في قوانين الجزء العملي من أجزاء الطبّ	822
the large quantity of thin pus	60	كُثْر: كثر الصديد الرقيق	823
a large quantity	16	كثرة	824
a surplus of blood	37	ـ الدم	825

Translation	Sec. No.	Maimonides	
much foul matter	60	ـ الوضر	826
gum tragacanth	63	كثيراء	827
to thicken	24	(كثف) كثّف	828
thickening the body	26	تكثيف: تكثيف البدن	829
		كرة ← لعب	
distress	30	كرب	830
		كرسنة ← دقيق	
trotters	69	كراع: أكارع(؟)	831
breaking up, fracture	16, 84	كسر	832
		مكسور ← علاج	
		منكسر ← جبر	
exposed	83	مكشوف	833
barley groats	15, 27, 33	كشك: كشك الشعير	834
		كلب ← عضّة	
		كلام ← قلّة	
kidneys	64	كلية: كلى	835
dose, quantity	17, 39, 60	كمية	836
quantities	2	كميات	837
to apply a hot compress	79	(كمد) كمّد	838
compacting, lividness, lividity	46, 60, 61	كمودة	839
frankincense	62	كندر	840
		كهرباء ← قرص	
elderly	10	كهل: كهول	841

Translation	Sec. No.	Maimonides	
to cauterize	56	كوى	842
cauterization	56, 76	كيّ	843
cauterization	76	الكيّ: الكيّ بالنار	844
		كاو ← أدوية	
capsule	49	كيس	845
quality	39, 60	كيفية	846
qualities	2	كيفيات	847
chyme	85	كيموس ← أدوية	848
dress him	34	لبس: لبّسه ثيابه	849
milk	35, 66	لبن	850
milk of a donkey	35	- الأتن	851
milk of a goat	35	- الماعز	852
milk of women	35	- النساء	853
cooled sour milk	27	اللبن: اللبن المحيض المبرّد	854
milk that is sucked from the breast	35	- المصّ من الثديين	855
milk	35	ألبان	856
to adhere to, to become congested	19, 42	لجّ	857
to make the flesh grow	62	(لحم)لحّم	858
		ألحم ← أدوية	
to close up	79	التحم	859

Translation	Sec. No.	Maimonides	
flesh	48, 55, 58, 59, 76	لحم ← أدوية، ذوبان، سوء، إنبات، انتبات، مواضع	860
the flesh of the breasts	48	ـ الثديين	861
meat of a kid	16	ـ الجدي	862
meat of a lamb	16	ـ الحمل	863
the flesh of the neck, the armpits, and the groin	48	ـ العنق والإبط والأربية	864
soft flesh	31	اللحم: اللحم الرخو	865
		لحمي ← أعضاء	
closing up	58	إلحام	866
formation of flesh	57	التحام	867
goatsbeard	63	لحية التيس	868
pleasant	26	لذيذ	869
to irritate	12	لذع	870
burning, pungency	31, 83	لَذْع	871
		لذّاع ← خلط	
		تلذيع ← أدوية	
irritation, inflammation	15, 63	تلذُّع	872
viscous, glutinous	3, 19, 61, 68	لزج ← خلط، أخلاط	873
viscosity	15, 33, 69, 85	لزوجة	874
to put	79	لزق	875
to adhere to	67	لزم	876

Translation	Sec. No.	Maimonides	
to adhere, to be cohesive	64, 74	لصق	877
adhering	75	لاصق	878
salve	85	لطوخ	879
to refine, to give a refining treatment	19, 33, 70	(لطف) لطّف ← أدوية	880
fine, softening	15, 32, 44, 82, 83, 85	لطيف ← جوهر، أدوية	881
finer	3	ألطف ← رطوبة	882
refinement, lightening	3, 20	تلطيف	883
thinning the humors	26	ـ الأخلاط	884
refining, dissolving, and evacuating the [crude] humors	23	ـ الأخلاط وتقطيعها واستفراغها	885
gentleness	84	تلطُّف	886
thinning out	20	ملطّف ← دواء	887
playing with the [small] ball	5	لَعب: اللعب بالكرّة	888
mucilage of fenugreek	15	لعاب: لعاب الحلبة	889
to bandage	84	لفّ: لفّ من اللفائف	890
		لفافة: لفائف ← لفّ	
lying on one's back	71	استلقاء: الاستلقاء على القفاء	891
burning heat	60	لهب	892
flare-up	30	التهاب	893
		لون ← حمرة	
fibers	57	ليف	894

Translation	Sec. No.	Maimonides	
to soften	47	(لين) ليّن	895
		لِين ← منديل	
soft	7	ليّن ← أدوية، أغذية	896
		متين ← دم	
		محيض ← لبن	
the length of the disease	20	مدّة: مدّة المرض	897
material, matter	16, 21, 36, 45, 53, 56, 58, 59, 68, 72, 73, 82	مادّة ← جَذْب، اجتذاب، انصباب، إمالة	898
matters	9, 45	موادّ	899
straightening the limb	84	تمديد: تمديد العضو	900
myrrh	62	مرّ	901
bile	11	مرّة	902
biles	30	أمرار	903
bile	40	مرار	904
		مراري ← بدن	
		امرأة: نساء ← لبن	
esophagus	64	مريء	905
digestion	35	(مرأ) استمراء	906
litharge	62	مرتك	907
to rub	32, 34, 83	مرخ	908
rubbing	7, 31, 34	مَرْخ	909
rubbing	8, 82	تمرُّخ	910

Translation	Sec. No.	Maimonides	
illness, disease, ailment	0, 1, 14, 16, 17, 18, 20, 50	مرض ← مدّة	911
illnesses	11	أمراض	912
patient	20, 34, 35, 37, 85	مريض	913
temperament	13, 16, 17, 19, 29, 40, 51, 52, 58, 59, 66, 71	مزاج ← سوء	914
a hot temperament	17, 40	المزاج: المزاج الحارّ	915
temperaments	4	أمزجة	916
temperaments	10	مزاجات	917
mixed	16, 50, 66	ممزوج ← خلّ، شراب	918
to rub (off)	23, 34	مسح	919
to contain	76	مسك	920
to keep in the mouth	64	- في الفم	921
to be contained	76	امتسك	922
abstention from	13	إمساك عن	923
		مصّ ← لبن	
stomach	3, 4, 6, 16, 17, 19, 20, 23, 27, 31, 32, 35, 45, 64	معدة ← فم	924
stomachs	24, 31	معد	925
		ماعز ← لبن	
gut	79	معاء	926

Translation	Sec. No.	Maimonides	
the jejunum	77	المعاء: المعاء الصائم	927
intestines	6, 31, 32, 64, 80	أمعاء	928
the small intestines	77	الأمعاء: الأمعاء الدقاق	929
the large intestines	64, 77	ـ الغلاظ	930
the rectum	17	ـ المنتصبة	931
to be saturated	34	(ملأ) امتلأ	932
congestion, overfilling	20, 22	امتلاء	933
		امتلائي ← حمّيات	
overfilled, replete	20, 45	ممتلئ	934
salt	50	ملح	935
salty	12	مالح ← مياه	936
water	16, 34, 50, 66, 72	ماء ← خلّ، شرب، إصلاح	937
lye from ashes	50	ـ الرماد	938
pomegranate juice	33	ـ الرمّان	939
sugar water	66	ـ السكّر	940
barley gruel	35, 72	ـ الشعير	941
hydromel, honey water	23, 66, 72	ـ العسل	942
chicory juice	15	ـ الهندباء	943
cold water	27, 30, 31, 34	الماء: الماء البارد	944
hot water	79, 82	ـ الحارّ	945
moderately hot water	34	ـ الحارّ المعتدل الحرارة	946

Translation	Sec. No.	Maimonides	
sweet water	66	ـ العذب	947
sweet water that is moderately hot	34	ـ العذب المعتدل الحرارة	948
sweet, lukewarm water	7	ـ العذب الفاتر	949
sweet waters	12	المياه: المياه العذبة	950
sweet, lukewarm waters	11	ـ العذبة الفاطرة	951
astringent waters	11, 31	ـ القابضة	952
salty waters	12	ـ المالحة	953
		مائي ← شراب	
liquid	62	ماعي	954
probe	49	ميل	955
diverting	73	إمالة	956
diversion of the matter	21	ـ المادّة	957
		مائل ← جزء	
		ناردين ← دُهن	
to be grown	58	نبت	958
		أنبت ← أدوية	
growing flesh, the growth of flesh	57, 59, 74	إنبات: إنبات اللحم	959
the growth of the flesh	57	انتبات: انتبات اللحم	960
coppersmith	12	نحّاس	961
		نحيف ← بدن، أبدان	
the nostril from which the blood flows	73	منخر: المنخر المرعوف	962
to puncture	82	نخس	963

Translation	Sec. No.	Maimonides	
puncture	83	نخسة	964
rough linen cloths	23	منديل: مناديل خشنة	965
moist towels	34	- رطبة	966
linen cloths of moderate softness or roughness	23	- متوسّطة في اللين والخشونة	967
bleeding	76	نزف	968
		← دواء	
hemorrhage	76, 77, 79	- الدم	969
to go down	71	نزل	970
catarrh	69, 70	نزل	971
catarrh	31, 70	نزلة	972
		← إنضاج	
a hot defluxion	70	- حارّة	973
catarrhs	71	نزلات	974
accommodation	34	منزل	975
the mutual proportion	1	تناسب	976
		متناسب ← أسباب	
to cut with a saw	55	نشر: نشر بمنشار	977
		منشار ← نشر	
to dry	59	(نشف) نشّف	978
starch	63, 71, 72	نشاء	979
to be concocted	31, 33, 38	نضج	980
to concoct	23	أنضج	981
ripeness, concoction	30, 32	نَضْج	982

Translation	Sec. No.	Maimonides	
concocted	37	نضيج	983
concocting, concoction	7, 23, 25, 33	إنضاج	984
the concoction of the bodily humors	3	ـ أخلاط البدن	985
ripening the catarrh	70	ـ النزلة	986
		منضج ← دواء	
fomentations	25, 54	نطول: نطولات	987
fomentation	82	تنطيل	988
to pay attention to	2	نظر: نظر نصب العين	989
to cleanse	72	(نظف) نظّف	990
to revive, invigorate	16, 30	(نعش) أنعش	991
reviving the strength	16, 33	إنعاش	992
to become inflated	79	(نفخ) انتفخ	993
		نافخ ← ريح	
being absorbed, getting through	16, 64	نفوذ	994
penetration	16	تنفيذ	995
		نفساني ← روح	
		نَفَس ← ضيق	
inhalation	3, 64	تنفُّس ← آلات، إصلاح	996
to expel, remove	10, 11, 16, 23	نفض	997
		منفّط ← أدوية	
to be beneficial	18	نفع	998

Translation	Sec. No.	Maimonides	
to benefit	34	انتفع	999
useful purpose	48	منفعة	1000
useful, beneficial	15, 23, 36, 37, 71	نافع	1001
decreasing	3	نَقْص	1002
deficient	8	نقيص	1003
too small, deficient	3, 20	ناقص	1004
to cleanse	17	(نقي) نقّى	1005
cleansing	19	نقاء	1006
free from	3	النقاء: النقاء من	1007
evacuating the residues that have accumulated in the body	2	تنقية: تنقية الفضول المجتمعة في البدن	1008
the bite of the poisonous animal	58	نهشة: نهشة الحيوان المسموم	1009
to be emaciated, to emaciate	13, 16, 35	نهك	1010
weakening most of all	18	أنهك	1011
climax	46, 48	انتهاء	1012
attack	20	نوبة ← فتور	1013
a fever attack	20	ـ الحمّى	1014
fire	21	نار ← كيّ	1015
lime	76	نورة	1016
to put to sleep	80	(نوم) نوّم	1017
sleep(ing), nap	12, 23, 29, 69	نوم	1018

Translation	Sec. No.	Maimonides	
soft-boiled eggs	11	نيمرشت	1019
old age	2	هرم	1020
to shake softly	80	هزّ: هزّهزًا رفيقا	1021
to make lean	12	(هزل) أهزل	1022
		هزال ← بدن	
digestion	6, 16	هَضْم	1023
digestion	6, 16	انهضام ← سريع	1024
anxiety	29, 31	همّ	1025
		هندباء ← شراب، ماء	
air	3, 8, 34, 40, 79	هواء ← إصلاح	1026
hot air	80	- حارّ	1027
cold air	27	الهواء: الهواء البارد	1028
hot air	31	- الحارّ	1029
airs	14	أهوية	1030
to trigger	15	(هاج) أهاج	1031
sinew	55	وتر	1032
pain	15, 25	وجع	1033
severe pain	82	- شديد	1034
		أوجاع ← تسكين	
		وخم ← تخم	
burned seashells	65	ودع: الودع المحروق	1035
		ورد ← دهن	

Translation	Sec. No.	Maimonides	
to suffer from a tumor	15	ورم	1036
tumor	20, 23, 24, 30, 31, 45, 46, 52, 78, 82	وَرَمٌ	1037
a hot tumor	15	ـ حارّ	1038
a hot tumor	46	الورم: الورم الحارّ	1039
a soft swelling	50	ـ الرخو	1040
a hard tumor	47	ـ الصلب	1041
tumors	51	أورام	1042
affected by a tumor	20	وارم	1043
		وسخ ← وضر	
		وسط ← متعقّد(؟)	
moderate	31, 51	متوسّط ← منديل	1044
to widen	82	(وسع) وسّع	1045
wideness	24	سعة	1046
widening the ulcer	58	توسيع: توسيع القرحة	1047
eating abundant food	12	توسُّع: التوسّع في الغذاء	1048
a large quantity of foods and smells	16	ـ:التوسّع في الأغذية والأرائح	1049
putting together	84	اتّصال ← ردّ، تفرّق	1050
the joining together of the parts of the ulcer	67	ـ أجزاء القرحة	1051
foul matter, foulness	61, 65, 66, 83	وضر	1052

Translation	Sec. No.	Maimonides	
foul matter	59	الوضر: الوضر والوسخ ← كثرة	1053
spot, location, situation, site	7, 17, 55, 57, 62, 64, 76, 82, 84	موضع ← برد، حرارة	1054
very fleshy areas	77	مواضع: المواضع الكثيرة اللحم	1055
		وطئ ← فراش	
to produce	11, 85	(ولد): ولّد	1056
to arise, to originate, to be produced	31, 33, 40, 52, 59, 65	تولّد	1057
generation, production	14, 85	تولُّد	1058
producing	16, 48, 52, 85	مولّد ← أدوية	1059
originating	13	متولّد	1060
the heat of poisons	31	وهج: وهج السموم	1061
dryness	10, 30, 58, 65	يُبْس	1062
dry	13, 60	يابس ← بدن، أبدان، سوء	1063
arm	73	يد	1064
hands	7	أيدي	1065
hands	68, 80	يدان ← شدّ، عروق	1066

Notes to the English Translation

1. The term *asbāb* seems to be used in the sense of material causes, or even, by extension (as in Sufi texts), the materials themselves that are the presumed causes; here it refers to the correct proportions of the material elements, which proportions cause the symptoms that we identify as disease.

2. Lit., "The more it changes with changes in the air."

3. See the supplement.

4. In *Guide of the Perplexed* 3.25 (ed. and trans. Pines, 503), Maimonides speaks about acts that exercise the body, such as ball games, wrestling, boxing, and suspension of breathing (*ḥaṣr al-nafas*). See as well his *Medical Aphorisms* 18.2, 3. For more on exercise, see Maimonides' *Medical Aphorisms* 18; *Regimen of Health* 1.7; *Medical Aphorisms* 18.13; *On Asthma* 5.5; and Bos, "Preservation of Health," 220–22.

5. The idea of the existence of more than one digestion returns in his *On Hemorrhoids* 1.1 and in *Regimen of Health* 1.2, where Maimonides speaks about three digestions: the first in the stomach, the second in the liver, and the third in the other organs, which the nutriments reach via the veins. This concept ultimately goes back to Galen, who also explained the physiology of nutrition in terms of three orders of digestion: the first coction taking place in the stomach, the second in the liver—the major nutritive organ where the food is turned into blood—and the third in the rest of the organs, which the nutriments reach via the veins. See Galen, *In Hippocratis Librum de alimento commentarius* 2.3 (ed. Kühn, 15:234–35); and *De bonis malisque sucis* 5.17–18 (ed. Helmreich, 411).

6. For massage, see section 23 below; Maimonides, *On Asthma* 10.5–6 (ed. and trans. Bos, 53–54); and Bos, "Preservation of Health," 232. By "quantity" Maimonides means both the number of massage movements per unit time and probably also the pressure applied. We would call this intensity, but we prefer the literal translation "quantity" because in medieval thought/physics, *quantitas* and *intensitas* are two very different concepts.

7. Lit., ejects.

8. Lit., ejects.

9. See Maimonides, *Regimen of Health* 1.12: "The excellent foods upon which every one who desires to stay healthy should rely are well-prepared wheat bread . . . the meat of the chicken . . . and the yolk from chicken eggs. By well-prepared bread, I mean that it should be made from fully ripened wheat, after the

superfluous moistures have been dried out, but which is not so old that it begins to spoil. The bread should be made of coarsely ground grain, meaning that it should not be peeled and that its bran should not be removed through sifting. It should be clearly raised, salted, and well worked during kneading and baked in an oven. This is well-prepared bread according to the physicians; it is the best of foods."

10. See Maimonides, *Medical Aphorisms* 20.71: "The testicles of roosters provide extremely good nourishment and are the best food one can give to emaciated people and convalescents. All testicles of living creatures are hot and moist and clearly help sexual potency" (derived from al-Tamīmī).

11. "soft-boiled eggs" (*nīmrisht*; *i.e.*, *nīmbirisht*): See Vullers, *Lexicon persico-latinum*, 2:1393: "ovum sorbile, semicoctum"; and Maimonides, *Medical Aphorisms* 23.105: "An egg that has been boiled for a long time is called hard-[boiled]; the one that reaches moderate thickness from boiling is called moderately boiled and is the same as a soft-boiled egg. *De alimentorum facultatibus* 3."

12. "rockfish" (*al-samak al-raḍrāḍī*): Lit., "fish that lives amongst the pebbles"; for "rockfish" (*al-samak al-sakhūr* or *al-samak al-sakhrī*), see *Medical Aphorisms* 8.25, and especially 17.25: "Rockfish is rapidly digested. Together with its high digestibility, it is extremely good and beneficial to preserve the health of a human body since it produces blood of an intermediate consistency, neither thin and fine nor thick. *De alimentorum* [*facultatibus*] 3." On fish in general, see *Regimen of Health* 1.17.

13. Cf. Maimonides, *On Asthma* 9.11 (ed. Bos, 47): "Sharp foods produce yellow bile and are all bad."

14. Cf. Maimonides, *On Asthma* 3.3 (ed. Bos, 13): "One should avoid . . . everything which sends vapors to the brain."

15. Cf. Maimonides, *On the Regimen of Health* 1.7: "According to the physicians, not every movement is exercise. What they call exercise is a strong or fast movement or a combination of both, that is, a vigorous movement whereby breathing changes and a person begins to breathe heavily. Whatever exceeds this is exertion; that is to say, very strong exercise is called exertion. Not everyone can tolerate exertion nor does he need it. It is nevertheless better in the preservation of health than the neglect of exercise."

16. Cf. Maimonides, *Medical Aphorisms* 7.12 (ed. Bos, 26): "A dissolution of the pneuma is either due to the movements of the soul such as intense joy, which is called exultation, or intense pleasure or intense fear, and similarly, anxiety and anger. As part of all these movements, Galen [also] mentioned pain and sleeplessness. But I say that these two should be counted separately because there is nothing stronger for the dissolution of the pneuma than pain, and after that, sleeplessness."

17. For the thickening effect of pitch or the pitch ointment called *dropax*, see Galen, *De sanitate tuenda* 6.8 (ed. Koch, 183) and *Galen's Hygiene* (trans. Green, 256): "And it is possible to see many of those who were formerly slender made stout by this remedy. . . ."

18. The pneuma, or spirit—a sort of very subtle and fine matter—was considered to be "the link between the material and spiritual nature of man"; Ullmann, *Islamic Medicine*, 62–63. See also Ibn Riḍwān, *On the Prevention of Bodily Ills* (ed. Dols). Conforming to the dominant theory in the Middle Ages,

Maimonides distinguished three different kinds of pneuma—natural, vital, and psychical—that were vital for the condition and functions of the body and soul. In *Regimen of Health* 4.1–2, Maimonides discusses this subject extensively in the context of air pollution and its detrimental influence on the pneumas, since these originate from the air one inhales; see as well *On Asthma* 2.1 (ed. Bos, n. 4 and p. 124); and Bos, "Preservation of Health," 225–28.

19. On the notion of the "innate heat," the force sustaining life in living beings, see Galen, *Parts of the Body* (trans. May, 1:50–53); cf. Maimonides, *On Asthma* 6.1.

20. For the connection between the inhalation of healthy air and a healthy body and mind, see note 18.

21. For the importance Maimonides attached to theoretical medical knowledge, see Bos, "Maimonides' Medical Works and Their Contributions," 248–55.

22. See *Medical Aphorisms* 9.69 (ed. Bos, 74): "When a tumor occurs in the liver, an extremely strict diet is required. No food is more beneficial to it than barley groats because they cleanse without biting. . . . *De methodo* [*medendi*] 13."

23. See *Medical Aphorisms* 9.71 (ed. Bos, 74–75): "Cultivated and wild chicory have a temperament dominated by some cold and are also somewhat bitter, and together they have a moderately astringent effect. Because of the presence of these two qualities, they are amongst the best remedies for the treatment of a hot, bad temperament of the liver. They are not as harmful for a cold, bad temperament as are cold, moist remedies without astringency and without bitterness. The reason for this is that they cool the liver moderately, and strengthen it through their astringency, and cleanse it through their bitterness. They are beneficial for the liver against a simple bad temperament and against one combined with [serous] matter, for when they are mixed with honey, they make those serous moistures and other moistures flow and descend [from the body]. *Mayāmir* 8."

24. See Maimonides, *On Asthma* 3.1 (ed. Bos, 13), where Maimonides warns against adding honey and sugar to different kinds of bread and thereby causing "great harm to the liver."

25. See *Medical Aphorisms* 9.69 (ed. Bos, 74), where Maimonides warns in the name of Galen against treating a tumor in the liver "with pomegranate juice or apples or other astringent things," because then "the openings of the vessels contract" and the bile cannot be evacuated.

26. "The . . . fat": Cf. supplement.

27. See Langermann, "L'oeuvre médicale de Maïmonide," 285–87, on Maimonides' emphasis that treatment should always be gradual.

28. Cf. Maimonides, *Medical Aphorisms* 20.24: "Wine mixed with an equal amount of water heats the whole body and rapidly moves to all its limbs. It ameliorates and improves the humors of the body by balancing their temperament and by evacuating bad humors. *In Hippocratis Aphorismos commentarius* 7."

29. Cf. Maimonides, *Regimen of Health* 1.4: "The view of this servant regarding the determination of the quantity of food for someone who wishes to preserve his health is that he should eat, when the weather is temperate, a quantity that does not distend his stomach, does not burden it, and does not fail to digest."

30. Cf. Maimonides, *Medical Aphorisms* 20.19: "The best meat of land animals is pork; then comes the meat of kids, and then that of calves. The meat of lambs is moist, sticky, and slimy. As for the meat of other land animals, I recommend that anyone who cares about keeping his humors in a healthy condition avoid eating it. *De bonis malisque sucis*"; ibid., 25.10: "In his treatise *De bonis malisque sucis*, he holds that pork excels any other [kind of] praiseworthy food. After that [in excellence] comes kid's meat; and after that, calf's meat; and after that, lamb's meat."

31. Cf. Maimonides, *On Asthma* 7.3 (ed. Bos, 34–35): "Regarding water, most people know that if one drinks it with a meal, it keeps the food crude because it forms a barrier between it and the stomach so that the food floats and is badly digested. But if one has a fixed habit of doing so, one should take as little as possible and delay taking it as long as possible. The best time to drink water is about two hours after a meal. One should select water that is sweet, pure, light, free from any change in odor and is drawn on the same day from running water."

32. Perhaps read: "when the stomach hurts, the head shares [the pain]."

33. The diagnostic concept implied in this discussion is that of sympathy. For an extensive discussion of the subject, see Siegel, *Galen's System of Physiology*, 360–82.

34. This seems to be the paragraph of miscellanea: in addition to the factors already mentioned, which are discussed in the books, the physician may notice something in the particular patient that he is treating indicating that something else should be added, or that one of the recommended treatments is not working and should be stopped. Almost all of the factors discussed in this chapter can be found somewhere in Galen, but virtually none in the form given here, and not in this catechism.

35. "drastic": Lit., harsh.

36. "hiera picra": See Ullmann, *Medizin im Islam*, 296, s.v. "*iyārağ fīkrā*": a compound medicine with aloe as the main ingredient; and Galen, *De compositione medicamentorum secundum locos* 8.2 (ed. Kühn, 13:129): "the so-called pikron remedy consisting of 100 drachmas of aloe and 6 drachmas of other drugs." For its composition see the next note.

37. Cf. Maimonides, *Medical Aphorisms* 9.45 (ed. Bos, 68): "If someone's food often putrefies in his stomach, he benefits from emesis before eating and from drinking sweet wine. He should eat those foods that are not quickly putrefied and should accustom himself to relieving the bowels from time to time with those substances that have a moderately relieving force, such as [hiera] picra. But if one neglects the bad humor, healing is difficult. *De sanitate tuenda* 6"; ibid., 9:46 (ed. Bos, 69): "But if bad humors have entered into the substance of the coats of the stomach, the best kind of treatment is their evacuation with aloe or hiera [picra] ingested with water. The composition of the hiera [picra] is six *mithqāls* each of Chinese cinnamon, Indian nard, saffron, asarabacca, mastic and balsam wood, and one hundred *mithqāls* of aloe. The hiera should be prepared in two ways, with washed aloe and unwashed aloe. The washed type strengthens the stomach more, while the unwashed type is a stronger laxative. *De methodo medendi* 7."

38. Cf. Langermann, "*L'oeuvre médicale de Maïmonide*," 275–302.

39. But cf. Maimonides, *Medical Aphorisms* 10.68 (ed. Bos, 18): "Food [intake] is the worst thing for anyone whose fever is caused by overfilling, or an obstruction, or an [inflamed] tumor, or putrefaction of the humors. Do not feed them—not even when the [fever] attack abates. For anyone whose fever is caused by sleeplessness or worry or anxiety or emotions, abstention from food is the worst. One may feed them at any time during the attack of the fever, but especially when it abates. *De methodo* [*medendi*] 10."

40. I.e., that one has to draw superfluities by exercising or otherwise activating the opposite direction.

41. I.e., uncooked. The point of the message is to "cook" the humor—to induce digestion by means of heating.

42. Alternative reading: "do not bring about."

43. Cf. Maimonides, *Medical Aphorisms* 21.15: "The most appropriate substance for expectorating the thick humors is hydromel and, for the viscous humors, oxymel. The second [best], after hydromel, is barley gruel, and after barley gruel, sweet wine. *In Hippocratis De acutorum morborum* [*victu*] *commentarius* 3."

44. Cf. *Medical Aphorisms* 10.66, where for patients suffering from ephemeral fever, Maimonides (following Galen) recommends using oxymel and barley gruel to open the obstructions caused by bad humors.

45. "If . . . winds": Cf. the supplement.

46. For the vital role of the pneuma, which is basically refined air necessary for the proper functioning of both the body and the mind, see Maimonides, *Regimen of Health* 4.1; Bos, "Preservation of Health," 225–26; and Langermann, "*Al-Ṭayyib on Spirit and Soul*," 149–58.

47. Cf. the supplement.

48. For the importance of dyscrasia (bad temperament) in the treatment of any disease, see Maimonides, *Medical Aphorisms* 3.67 (ed. Bos, 50): "The most important and most dangerous element in treating diseases is a bad temperament. This is so because the temperament is the most important of the kinds [of things existing] in nature. *De methodo medendi* 7."

49. See section 33 below; see as well Maimonides, *Medical Aphorisms* 19.11: "The best thing for people from whose bodies a vaporous superfluity dissolves while they are healthy is bathing in sweet water. If you prevent these and their like from bathing, they develop a fever. . . . *De methodo medendi* 8."

50. "When . . . postponed": Cf. the supplement.

51. Cf. Maimonides, *Medical Aphorisms* 10.67 (ed. Bos, 18): "No fever develops from an indigestion in which the food turns sour. But fever can occur from an indigestion in which the food becomes gaseous. If someone develops a fever from such an indigestion and also suffers from diarrhea, and if you see that the evacuated matter consists only of that which was corrupted, let the patient bathe and feed him during the abatement of the first attack and take care to strengthen his stomach. But if you see that the earlier or later evacuation was so extreme that it wears out the strength [of the patient], the best thing [to do] is to feed him without having him go to the bathhouse. *De methodo* [*medendi*] 8."

52. Cf. Maimonides, *Medical Aphorisms* 9.47 (ed. Bos, 69): "[In the case of] indigestion associated with diarrhea so severe that it harms one's strength, one

should feed oneself with things that are astringent more and more. Often diarrhea goes with loss of appetite. In this case the patient should take a stomachic of quinces and the like. However, if indigestion is associated with constipation and either with a fever arising from the indigestion or without a fever, and if the spoiled food is in the upper parts of the abdomen, one should bring it down with the various kinds of pepper and the like. When the spoiled food has gone down, one should expel it either with a suppository or with an enema of honey, water, and olive oil. . . . If the patient suffers from flatulence, administer to him an enema of olive oil in which rue was cooked or in which seeds which disperse the winds, such as cumin, caraway, celery seed, and the like, were cooked. *De methodo medendi* 8."

53. For the subject of blood fever and its treatment, see *Ibn al-Jazzār, Zād al-musāfir wa-qūṭ al-ḥāḍir* (ed. Bos, 16–18, 120–27); and Langermann, "Synochous Fever," 175–98.

54. Cf. the supplement.

55. Cf. the supplement.

56. Cf. the supplement.

57. Cf. the supplement.

58. Cf. the supplement.

59. Cf. the supplement.

60. Cf. the supplement.

61. Cf. the supplement.

62. Cf. the supplement.

63. "If . . . appropriately": Cf. the supplement.

64. I.e., the ordering of the medications: they ought to be given in a certain order, progressing from weaker to stronger.

65. Cf. De Lacy, "Galen's Concept of Continuity."

66. On the theriac, a medical concoction made of opium, flesh of viper, and a large number of other ingredients, originally designed as an antidote against snake venom but eventually used as a preventative panacea, see Ullmann, *Medizin im Islam*, 321; Richter-Bernburg, "*De Theriaca ad Pisonem*," 115–17; and Fellmann, *Aqrābādhīn al-Qalānisī*, 274–78.

67. Maimonides, *Sharḥ asmāʾ al-ʿuqqār* (trans. Rosner, *Glossary of Drug Names*), no. 175, mentions the different kinds of earth.

68. For the composition of this dish, see al-Baghdadi, *Baghdad Cookery Book* (trans. Arberry), 71.

69. Cf. the supplement.

70. Cf. al-Zahrāwī (Albucasis), *On Surgery and Instruments* (ed. and trans. Spink and Lewis), 548: "You should know that if the injured bowel be the large intestine it will heal more easily, while if it be the small its healing is more difficult. But the part of the bowel called the jejunum is not likely to recover from an injury to it at all; and that is due to the number and size of arteries in it and to the thinness of its structure and to its nearly approaching a nerve in its character."

71. Cf. al-Zahrāwī (Albucasis), *On Surgery and Instruments* (ed. and trans. Spink and Lewis), 548: "If what protrudes from the wound be omentum, and you found it freshly protruding, then reduce it as you do intestine. But if some time has lapsed and it has turned green or black, you should ligate it above the site

where it has gone black lest there be a haemorrhage; for the omentum contains veins and arteries. Then cut off what is below the ligature and leave the end of the thread hanging out from the lower part of the wound, so that it will be easy for you to draw upon it and extract it as the omentum falls away and the wound suppurates."

72. Cf. al-Zahrāwī (Albucasis), *On Surgery and Instruments* (ed. and trans. Spink and Lewis), 536: "I begin then by saying that when the wound is small, and a piece of intestine protrudes from it, rendering the reduction of it difficult, the difficulty will be due to one of two causes: either on account of the smallness of the fissure, as we said; or because the gut has become inflated on account of the coldness of the air. If the latter case obtains, the gut must be warmed by fomenting with a sponge or piece of cloth bathed with warm water. . . . Wine that is somewhat astringent may do; in bringing down the swelling it is better than plain water."

73. cut it: Perhaps "operate upon it"?

74. That is, one should draw analogies between whatever Maimonides has written here and the specific case that the physician (the reader) may be treating, in order to arrive at the proper treatment.

75. Maimonides probably means that one should avoid taking hopeless cases, so that one will not be called a bad physician, in line with the idea that the physician will be blamed even though there was nothing he could do.

76. Note, however, that Maimonides himself did not always stick to this rule, but on occasion tried novel medicines that he himself composed, as he admits in some of his medical writings. See Bos, "Maimonides' Medical Works and Their Contributions," 256–57.

77. That is, Maimonides, *Medical Aphorisms*, book 13 (ed. Bos, 40–51).

Bibliographies

Editions and Translations of Works by Moses Maimonides

Barzel, Uriel S., ed. and trans. *The Art of Cure: Extracts from Galen*. Foreword by Fred Rosner; bibliography by Jacob I. Dienstag. Haifa: Maimonides Research Institute, 1992.

Bos, Gerrit, ed. and trans. *Commentary on Hippocrates' Aphorisms*. Provo, Utah: Brigham Young University Press, forthcoming.

———. *Medical Aphorisms*. 5 vols. Provo, Utah: Brigham Young University Press, 2004–.

———, ed. and trans. *On Asthma*. Provo, Utah: Brigham Young University Press, 2002.

———. *On the Elucidation of Some Symptoms and the Response to Them*. Provo, Utah: Brigham Young University Press, forthcoming.

Bos, Gerrit, ed. and trans., and Michael R. McVaugh, ed. *On Asthma, Volume 2*. Provo, Utah: Brigham Young University Press, 2008.

Bos, Gerrit, ed. and trans., and Charles Burnett, ed. *On Coitus*. Provo, Utah: Brigham Young University Press, forthcoming.

Bos, Gerrit, ed. and trans., and Michael R. McVaugh, ed. *On Hemorrhoids*. Provo, Utah: Brigham Young University Press, 2012.

———. *On Poisons and the Protection against Lethal Drugs*. Provo, Utah: Brigham Young University Press, 2009.

———. *The Regimen of Health*. Provo, Utah: Brigham Young University Press, forthcoming.

Meyerhof, Max, ed. and trans. *Sharḥ asmāʾ al-ʿuqqār* (*L'explication des noms des drogues*): *Un glossaire de matière médicale composé par Maïmonide*. Cairo: L'Institut français d'archéologie orientale, 1940. (See also Rosner's translation below.)

Pines, Shlomo, ed. and trans. *The Guide of the Perplexed*. 2 vols. Chicago: University of Chicago Press, 1963.

Rosner, Fred, trans. *Moses Maimonides' Glossary of Drug Names*. Haifa: Maimonides Research Institute, 1995. (See also Meyerhof's edition and translation above.)

General Bibliography

Ackermann, Hermann. "Moses Maimonides (1135–1204): Ärztliche Tätigkeit und medizinische Schriften." *Sudhoffs Archiv* 70, no. 1 (1986): 44–63.

Albucasis. *See* Zahrāwī, Abū al-Qāsim al-.

Algizar. *See* Ibn al-Jazzār al-Qayrawānī.

Anawati, Georges. *Essai de bibliographie avicennienne*. Cairo: Edition al-Maaref, 1950.

Avicenna. *See* Ibn Sīnā, Abū ʿAlī al-Ḥusayn.

Avishur, Yitzhak, ed. *Shivḥe ha-Rambam: Sippurim ʿamamiyim be-ʿAravit Yehudit uve-ʿIvrit meha-Mizraḥ umi-Tsefon Afriḳah*. Jerusalem: The Hebrew University Magnes Press, 1998.

Baghdadi, Muḥammad ibn al-Ḥasan al-. *A Baghdad Cookery Book (Kitāb al-ṭabīkh)*. Translated by A. J. Arberry. In *Medieval Arab Cookery: Essays and Translations*, edited by Maxime Rodinson, A. J. Arberry, and Charles Perry. Blackawton, Totnes, Devon, England: Prospect, 2001.

Baron, Salo Wittmyer. *A Social and Religious History of the Jews*. Vol. 8, *High Middle Ages, 500–1200: Philosophy and Science*. 2nd ed. New York: Columbia University Press, 1958.

Ben-Sasson, Menahem. "Maimonides in Egypt: The First Stage." *Maimonidean Studies* 2 (1991): 3–30.

Bos, Gerrit. "Maimonides on Medicinal Measures and Weights." *Aleph* 9, no. 2 (2009): 255–76.

———. "Maimonides on the Preservation of Health." *Journal of the Royal Asiatic Society*, 3rd ser., 4, no. 2 (1994): 213–35.

———. "Maimonides' Medical Works and Their Contributions to His Medical Biography." *Maimonidean Studies* 5 (2008): 244–48.

Cohen, Mark R. "Maimonides' Egypt." In *Moses Maimonides and His Time*, edited by Eric L. Ormsby, 21–34. Washington, DC: The Catholic University of America, 1989.

Davidson, Herbert A. "Maimonides' Putative Position as Official Head of the Egyptian Jewish Community." In *Ḥazon Naḥum: Studies Presented to Dr. Norman Lamm in Honor of His Seventieth Birthday*, edited by Yaakov Elman and Jeffrey S. Gurock, 115–28. New York: Yeshiva University Press, 1997.

———. *Moses Maimonides: The Man and His Works*. New York: Oxford University Press, 2005.

De Lacy, Phillip. "Galen's concept of continuity." *Greek, Roman, and Byzantine Studies* 20, no. 1 (1979): 355–69.

Dienstag, Jacob I. "Translators and Editors of Maimonides' Medical Works." In *Memorial Volume in Honor of Prof. Süssmann Muntner*, edited by Joshua O. Leibowitz, 95–135. Jerusalem: Israel Institute for the History of Medicine, 1983.

Encyclopaedia of Islam. New ed. 12 vols. Leiden: Brill, 1960–94.

Encyclopaedia Judaica. 16 vols. Jerusalem: Keter, 1971–72.

Fellmann, Irene. *Das* "Aqrābādhīn al-Qalānisī": *Quellenkritische und begriffsanalytische Untersuchungen zur arabisch-pharmazeutischen Literatur*. Beirut: Orient-Institut der Deutschen Morgenländischen Gesellschaft, 1986.

Friedenwald, Harry. *The Jews and Medicine: Essays*. 2 vols. 1944. Reprint, New York: Johns Hopkins University Press, 1967.

Friedman, M. "Ha-Rambam 'Raʾis al-Yahud' (Rosh ha-Yehudim) be-Miẓrayim." In *ʿAl pi ha-beʾer: Meḥkarim be-hagut Yehudit uve-maḥshevet ha-halakhah mugashim le-Yaʿakov Blidshtain [By the well: Studies in Jewish philosophy and halakhic thought presented to Gerald J. Blidstein]*, edited by Uri Ehrlich, Howard T. Kreisel, and Daniel J. Lasker, 413–35. Beer Shevaʿ, Israel: Universitat Ben-Gurion ba-Negev, 2007.

Galen. *Claudii Galeni opera omnia*. Edited by Karl Gottlob Kühn. 20 vols. 1821–33. Reprint, Hildesheim, Germany: Olms, 1964–67.

———. *De bonis malisque sucis*. Edited by Georg Helmreich. Corpus Medicorum Graecorum 5.4.2. Leipzig: Teubner, 1923.

———. *De sanitate tuenda*. Edited by Konrad Koch. Corpus Medicorum Graecorum 5.4.2. Leipzig: Teubner, 1923.

———. *Galen on the Usefulness of the Parts of the Body*. Translated by Margaret Tallmadge May. 2 vols. Ithaca, NY: Cornell University Press, 1968.

———. *A Translation of Galen's "Hygiene."* Translated by Robert M. Green. Introduction by Henry E. Sigerist. Springfield, IL: Thomas, 1951.

Goitein, Shelomoh D. "Ḥayyē ha-Rambam le-ʾOr Gilluyim ḥadashim min ha-genizah ha-ḳahirit." *Peraḳim* 4 (1966): 29–42.

———. "Moses Maimonides, Man of Action: A Revision of the Master's Biography in Light of the Geniza Documents." In *Hommage à Georges Vajda: Études d'histoire et de pensées juives*, edited by Gérard Nahon and Charles Touati, 155–67. Leuven, Belgium: Peeters, 1980.

Graetz, Heinrich. *Geschichte der Juden*. 11 vols. Leipzig: Leiner, 1890–1909.

Guillén Robles, Francisco. *Catálogo de los manuscritos árabes existentes en la Biblioteca Nacional de Madrid*. Madrid: Imprenta y Fundación de Manuel Tello, 1889.

Haly Abenrudian. *See* Ibn Riḍwān, Abu al-Ḥasan.

Ibn al-Jazzār al-Qayrawānī. *Ibn al-Jazzār on Fevers: A Critical Edition of "Zād al-musāfir wa-qūt al-ḥāḍir." Provisions for the Traveller and Nourishment for the Sedentary*. Bk. 7, chaps. 1–6. Edited and translated by Gerrit Bos. London: Kegan Paul International, 2000.

Ibn Riḍwān, Abu al-Ḥasan. *Medieval Islamic Medicine: Ibn Riḍwān's Treatise "On the Prevention of Bodily Ills in Egypt."* Edited and translated by Michael W. Dols. Berkeley and Los Angeles: University of California Press, 1984.

Ibn Sīnā, Abū ʿAlī al-Ḥusayn. *Kitāb al-qānūn fī al-ṭibb*. 5 bks. in 3 vols. 1877. Reprint, Beirut: Dār Ṣādir, n.d.

Kraemer, Joel L. "The Life of Moses ben Maimon." In *Judaism in Practice: From the Middle Ages through the Early Modern Period*, edited by Lawrence Fine, 413–28. Princeton, NJ: Princeton University Press, 2001.

———. *Maimonides: The Life and World of One of Civilization's Greatest Minds*. New York: Doubleday, 2008.

———. "Maimonides' Intellectual Milieu in Cairo." In *Maïmonide: Philosophe et savant (1138–1204)*, edited by Tony Lévy and Roshdi Rashid, 1–37. Leuven, Belgium: Peeters, 2004.

Langermann, Y. Tzvi. "Abū al-Faraj ibn al-Ṭayyib on Spirit and Soul." *Le Muséon* 122 (2009): 149–58.

———. "Another Andalusian Revolt? Ibn Rushd's Critique of al-Kindī's Pharmacological Computus." In *The Enterprise of Science in Islam: New Perspectives*, edited by Jan P. Hogendijk and Abdelhamid I. Sabra, 351–72. Cambridge, MA: MIT Press, 2003.

———. "Fūṣūl Mūsā: Maimonides' Method of Composition." *Maimonidean Studies* 5 (2008): 325–44.

———. "L'oeuvre médicale de Maïmonide: Un aperçu général." In *Maïmonide: Philosophe et savant (1138–1204)*, edited by Tony Lévy and Roshdi Rashid. Leuven, Belgium: Peeters, 2004.

———. "Maimonides on the Synochous Fever." *Israel Oriental Studies* 13 (1993): 175–98.

Leibowitz, Joshua O. "Maimonides: Der Mann und sein Werk: Formen der Weisheit." *Ariel* 40 (1976): 73–89.

Levinger, Jacob. "Was Maimonides 'Rais al-Yahud' in Egypt?" In *Studies in Maimonides*, edited by Isadore Twerski, 83–93. Cambridge, MA: Harvard University Press, 1990.

Lewis, Bernard. "Maimonides, Lionheart and Saladin." In *Erets-Israel: Archaeological, Historical, and Geographical Studies* 7, edited by M. Avi-Yonah, H. Z. Hirschberg, B. Mazar, and Y. Yadin, 70–75. Jerusalem: Israel Exploration Society, 1964.

McVaugh, Michael R. *The Rational Surgery of the Middle Ages*. Florence: SISMEL/Edizioni del Galluzzo, 2006.

Meyerhof, Max. "The Medical Work of Maimonides." In *Essays on Maimonides: An Octocentennial Volume*, edited by Salo Wittmayer Baron, 265–99. New York: Columbia University Press, 1941.

Papavramidou, Niki, and Helen Christopoulou-Aletra. "The Ancient Technique of 'Gastrorrhaphy.'" *Journal of Gastrointestinal Surgery* 13 (2009): 1345–50.

Richter-Bernburg, Lutz, trans. "Eine arabische Version der pseudogalenischen Schrift *De Theriaca ad Pisonem*." PhD diss., University of Göttingen, 1969.

Shailat, Isaac. *Iggerot ha-Rambam*. 2 vols. Jerusalem: Maʿaliyot, 1987–88.

Siegel, Rudolph E. *Galen's System of Physiology and Medicine: An Analysis of His Doctrines and Observations on Bloodflow, Respiration, Tumors and Internal Diseases*. New York: Karger, 1968.

Steinschneider, Moritz. *Die hebräischen Übersetzungen des Mittelalters und die Juden als Dolmetscher*. 1893. Reprint, Graz, Austria: Akademische Druck- und Verlagsanstalt, 1956.

Ullmann, Manfred. *Die Medizin im Islam*. Leiden: Brill, 1970.

———. *Islamic Medicine*. Translated by Jean Watt. Edinburgh: Edinburgh University Press, 1978.

Vullers, Johann August. *Lexicon persico-latinum etymologicum*. 2 vols. 1855–64. Reprint, Graz, Austria: Akademische Druck- und Verlagsanstalt, 1962.

Zahrāwī, Abū al-Qāsim al-. *Albucasis on Surgery and Instruments*. Edited and translated by Martin S. Spink and Geoffrey L. Lewis. Berkeley: University of California Press, 1973.

Subject Index to the English Translation

Note: The locators refer to section numbers, not page numbers; 0 refers to Maimonides' introduction.

About the Editors

GERRIT BOS is chair of the Martin Buber Institute for Jewish Studies at the University of Cologne. He is proficient in classical and Semitic languages and is widely published in the fields of Jewish studies, Islamic studies, Judeo-Arabic texts, and medieval Islamic science and medicine. In addition to preparing the Medical Works of Moses Maimonides, Professor Bos is involved with a series of medical-botanical Arabic-Hebrew-Romance synonym texts written in Hebrew characters, and is producing an edition of Ibn al-Jazzār's *Zād al-musāfir* (Viaticum). He is also studying the Hebrew medical terminology used by the major translators of the thirteenth century; a first analysis can be found in his *Novel Medical and General Hebrew Terminology from the 13th Century* (2 vols.). He received the Maurice Amado award for his work on Maimonides' medical texts and is a Member of Honor of the Argentinean Society for the History of Medicine.

TZVI LANGERMANN received his PhD in history of science from Harvard University. He teaches in the Department of Arabic, Bar Ilan University, Ramat Gan, Israel. Langermann publishes on a wide variety of topics related to science, philosophy, and religious thought in medieval Judaism and Islam. He recently edited a collection of essays, *Monotheism and Ethics*, published by Brill in 2011.

A Note on the Types

The English text of this book was set in BASKERVILLE, a typeface originally designed by John Baskerville (1706–1775), a British stonecutter, letter designer, typefounder, and printer. The Baskerville type is considered to be one of the first "transitional" faces—a deliberate move away from the "old style" of the Continental humanist printer. Its rounded letterforms presented a greater differentiation of thick and thin strokes, the serifs on the lowercase letters were more nearly horizontal, and the stress was nearer the vertical—all of which would later influence the "modern" style undertaken by Bodoni and Didot in the 1790s. Because of its high readability, particularly in long texts, the type was subsequently copied by all major typefoundries. (The original punches and matrices still survive today at Cambridge University Press.) This adaptation, designed by the Compugraphic Corporation in the 1960s, is a notable departure from other versions of the Baskerville typeface by its overall typographic evenness and lightness in color. To enhance its range, supplemental diacritics and ligatures were created in 1997 for exclusive use in this series.

The Arabic text was set in NASKH, designed by Thomas Milo (b. 1950), a pioneer of Arabic script research, typeface design, and smart font technology in the digital era. The Naskh calligraphic style arose in Baghdad during the tenth century and became very widespread and refined during the Ottoman period. It has been favored ever since for its clarity, elegance, and versatility. Milo designed and expanded this typeface during 1992–1995 at the request of Microsoft's Middle East Product Development Department and extended its typographic range even further in subsequent editions. Milo's designs pushed the existing typographic possibilities to their limits and led to the creation of a new generation of Arabic typefaces that allowed for a more authentic treatment of the script than had been possible since the advent of moveable type for Arabic.

BOOK DESIGN BY JONATHAN SALTZMAN

◆